Jr. Boom Academy

Conceived, constructed, and doodled
by the folks at The Wild Goose Company

Copyright 1992 • Revised 1999

the
Wild
Goose
company
Real Science — Real Fun!®

Greensboro, NC 888-621-1040

Table of Contents

Table of Contents

Table of Contents

Lab Safety

In every lab class there's always the danger you may expose yourself to injury. The chemicals and equipment you use and the way you use them are very important, not only for your safety but for the safety of those working around you. Please observe the following rules at all times. Failure to do so increases your risk of accident.

1. Goggles.

Goggles should always be worn when chemicals are being heated or mixed. This will protect your eyes from chemicals that spatter or explode. Running water should be available. If you happen to get some chemical in your eye, flush thoroughly with water for **15 minutes.** If irritation develops, contact a physician. Take this book and the bottle of the chemical with you the to the doctor's office.

2. Smelling Chemicals.

If you need to smell a chemical to identify it, hold it six inches away from your nose and wave your hand over the opening of the container toward your nose. This will "waft" some of the fumes toward your nose without exposing you to a large dose of anything really stinky or dangerous.

3. Chemical Contact With Skin.

Your kit contains protective gloves to wear whenever you are handling chemicals. If you do happen to spill a chemical on your skin, flush the area with water for **15 minutes.** If irritation develops, contact a physician. Take this book and the bottle of chemical with you to the doctor.

5. Proper Disposal of Poisons.

If a substance needs special disposal is used or formed during the experiments in this lab, the book will tell you. These must be handled according to the directions in the lab guide.

Lab Safety

6. No Eating or Drinking During the Lab.
When you eat or drink in the lab, you run the risk of internalizing poison. This is never done unless the lab calls for it. Make sure your hands and lab area are always clean when you're finished experimenting.

7. Horseplay Out.
Horseplay can lead to chemical spills, accidental fires, broken containers, damaged equipment, and injured people. Never throw anything to or at another person! Be careful where you put your hands and arms. No wrestling, punching, or shoving in the lab.

8. Fire.
Remember the rule: No adults in the room = no flames allowed! Get adult help with any fire that's not part of the lab. Know where to locate and how to use a fire extinguisher. If clothes are on fire; STOP, DROP, and ROLL!

9. Better Safe Than Sorry.
If you have questions, or if you are not sure how to handle a particular chemical, procedure, or part of an experiment, ask for help from an adult. If you don't feel comfortable doing something, then don't do it. If there is any concern upon chemical exposure, contact a physician. Take this book and the bottle of chemical with you.

Lab safety is important! I tell you this because I want you to have fun and be safe when you do the experiments in this book or whenever you're working in a lab. Have fun!

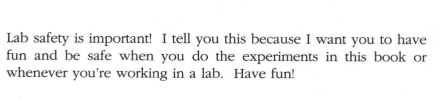

Acquiring the Tools of the Trade

Thermometer Reading

So What's Up?

Measure and record the temperature of water using in Celsius (°C), Fahrenheit (°F), and Kelvin (°K) thermometer scales.

"Why?" You Ask

Scientists use many different kinds of tools to help them study the world. One of the most common tools we use every day is the thermometer. The Fahrenheit thermometer is the conventional temperature scale used here in the United States. The Celsius (also called centigrade) scale is based on the metric system and is used by the rest of the world except for a few African countries. The Kelvin Scale is sometimes called the "absolute" temperature scale. Zero on the Kelvin Scale equals -273°C or -459.4°F. The Fahrenheit scale reads 32° for freezing and 212° for boiling; Celsius is 0° and 100° for the same readings. Generally, every one degree Celsius change is equal to 2.3 degrees Fahrenheit.

Materials and Ingredients

- 6 ice cubes
- 3 beakers
- 1 Celsius thermometer
- 1 container of hot water
- 1 container of room temperature water
- 1 Fahrenheit thermometer
- 1 pen
-    masking tape

Watch Out For . . .

The hot water needs to be handled carefully so nobody gets burned.

The Procedure

1. Tear off and stick one piece of tape on each of your three beakers.

2. Label the beakers A, B, and C, using the pen.

3. Put the six ice cubes into beaker A and fill the beaker half-full of room temperature water. Put it on your desk.

4. Fill beaker B half-full of room temperature water and put it next to beaker A.

5. Fill beaker C with hot water and carefully place the beaker next to the other two.

6. Place both thermometers in beaker A for 15 seconds. This will give the thermometer a chance to record an accurate reading. Record the temperature of the water in the data table on the next page in Celsius and Fahrenheit degrees.

7. Do the same for beakers B and C.

Thermometer Reading

Data and Observations

To calculate degrees Kelvin, add 273.15 to the Celsius reading.

Beaker	Celsius (°C)	Fahrenheit (°F)	Kelvin (°K)
A			
B			
C			

Questions for the Anklebiters

1. Your mother keeps complaining that the thermometer readings on the bank reader board are in Celsius. In fact, just this morning she asked you, "How are Fahrenheit thermometer scales different from Celsius and Kelvin?" You, being the child genius you are, explain this to your mom.

   _____

   _____

   _____

2. What is the temperature of the ice water in:

   a. Celsius? _____

   b. Fahrenheit? _____

   c. Kelvin? _____

3. How do the three temperature scales differ?

   _____

   _____

   _____

4. How much hotter was the hot water than the cool water in:

   a. Celsius? _____

   b. Fahrenheit? _____

   c. Kelvin?_____

Thermometer Reading

Ideas to Try

Heat water prior to class so it will be ready for the kids when they get there. Use a hot plate and large pan. To get things going, select one student and have him or her come to the front of the class. Tell the class this student will be a living thermometer. Follow these steps:

1. Blindfold the student and tell him or her to rely on sense of touch only. If he or she is apprehensive, you may want to explain that the experiment is completely safe.

2. Have three gallon-size pans set out in front of the student. The pan on the right should have very warm water, the pan in the middle should have room temperature water, and the pan on the left should contain ice water.

3. Have the student put his or her left hand in the left pan and his or her right hand in the right pan with the middle pan left untouched. Instruct the student not to remove his or her hands until instructed to do so.

4. Let the students hands soak in the water for 30 seconds. Take the student's right hand out of the water and put it in the pan in the middle. Ask him to describe the temperature of the water. The student will probably say it is cool. Put the student's right hand back in the original bucket, then, place his left hand in the middle pan and ask for a description of the temperature. The student will probably say the water is warm.

5. Take the blindfold off and show the student that he or she was describing the same pan of water. Explain to the class it's important we have an accurate and consistent way to measure temperature so that all collected information is understood and recorded in the same way.

Heating of Ice Water

So What's Up?
Determine the boiling point of a liquid by recording the temperature of the liquid as you heat it.

"Why?" You Ask
When a substance or liquid reaches the boiling point, the temperature will not increase any more. When the molecules move faster than a certain speed, they break free from the liquid. The molecules left behind remain at a more-or-less constant average speed. Because the temperature of the liquid is a measure of the average speed of the molecules, the temperature remains constant.

Materials and Ingredients
1 beaker, 400 ml
1 Celsius thermometer
1 hot plate
1 pair of goggles
1 spill apron
water

Watch Out For . . .
The hot plate can be very dangerous if left unattended or used without care. Always have an adult help you.

The Procedure
1. Put the goggles on your face and the spill apron around your body. Then add 250 ml of water to the beaker.
2. Put the beaker on the hot plate and plug that baby in. Needless to say, it's called a hot plate because it gets very hot.
3. Put the thermometer into the water and record the temperature every minute until the water has been boiling for five minutes.
4. Unplug the hot plate when you're finished and wait for all the equipment to cool down before you put it away.

Goggles

Spill Apron

Heating of Ice Water

Data and Observations

Record the temperature every minute until the water boils. Measure for five minutes after the water boils.

Time (minutes)	Temp. (celsius)	Time (minutes)	Temp. (celsius)
0		8	
1		9	
2		10	
3		11	
4		12	
5		13	
6		14	
7		15	

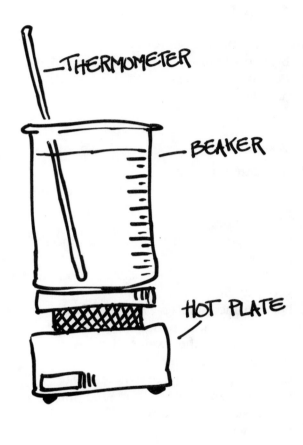

Questions for the Anklebiters

1. At what temperature did the water begin to boil?_____

2. Did the temperature level off after it reached boiling?_____

3. What is the boiling point of water?_____

Beakers & Balances

So What's Up?

Develop the skills and knowledge needed to read beaker and scale measurements.

"Why?" You Ask

Beakers and balances are two tools scientists use to record information about their experiments accurately. Beakers are used to measure volume and run chemical experiments while balances determine the mass of objects. These instruments are important because without an independent and consistent way to measure, scientists would not be able to communicate with one another, repeat experiments, or share data.

Materials and Ingredients

10 objects to weigh
3 beaker, 400 ml
2 beaker, 1000 ml
1 double-pan balance

The Procedure

1. Put the five beakers on a table. The markings on the side of the beakers are in milliliters. This is abbreviated "ml." It is one one-thousandth of a liter. When you record the volume of a container, you must always put the unit you are using. Practice filling and reading the beakers within your lab group. Fill the container to the following volumes: 10ml, 125 ml, 750 ml, 30 ml, and 950 ml. You may want to use different-sized beakers for the different measurements. Check each other to make sure you are reading the beaker properly.

2. There are many different kinds of balances. Your teacher will instruct you about how to use them. Record the weights of the ten different objects in the data table on the next page.

3. When it's time for the lab groups to come up and read the five beakers, record the measurements on the chart below.

Beaker #	1	2	3	4	5
Volume (ml)					

Beakers & Balances

Data and Observations

Weight Measurements

Object #	1	2	3	4	5
Mass (g)					
Object #	6	7	8	9	10
Mass (g)					

In the spaces below, draw beakers with the following volumes in them. Be sure to show the markings on the sides of the containers clearly and color in the volumes.

250 ml   700 ml   0 ml

Beakers & Balances

Ideas to Try

1. Find a metric recipe and have the kids measure the ingredients for cookies using graduated cylinders, beakers, and balances.

2. Have the first annual Measurement Relays. Break the class into teams of five kids and head into the gymnasium. At one end of the gym the kids line up. At the other end of the gym are the judges. They can be parents, the secretary, school psychologist, custodian, or whoever happens to wander in at the time. There are index cards with volumes or masses written on them (10 ml, 43 ml, etc.), a graduated cylinder, various beakers and a large beaker of water (use plastic beakers). At the sound of "Go!" the first kid from each team races down to the other end of the gym and flips over the first card. There is a volume written on that card. The kid picks up the beaker of water and fills the graduated cylinder or beaker, whichever is indicated on the card, to the exact amount. The judge verifies the measurement, the kid pours the water back into the beaker, and runs back to his or her team. The next member of the team zips down and repeats the procedure. Guaranteed pandemonium and a whole lot of fun!

Jr. Boom Academy

Graduated Cylinders

So What's Up?

Learn how to fill, read, and record measurements on a graduated cylinder.

"Why?" You Ask

Scientists sometimes need to measure the volume of a fluid very accurately. To accomplish this, they use a long, skinny container. It is called a "**graduated**" cylinder because it is marked off in even segments. When you pour a liquid, like water, into a graduated cylinder, the water molecules stick to the sides of the container causing a dip, or **meniscus** (men-iss-cuss), to form. To measure the volume in the cylinder, you have to look at the center point of the meniscus at eye level and read the graduation in milliliters. When you record the volume of a container, you must always put the unit of measurement you are using. In this case it will always be "ml."

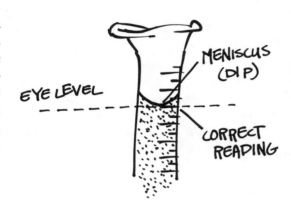

Materials and Ingredients

    1 beaker, 400 ml
    1 graduated cylinder
       water

The Procedure

1. Practice filling and reading the graduated cylinder for the other members in your group. Fill the container to the following volumes: 10 ml, 25 ml, 26 ml, 30 ml, and 50 ml. Check each other to make sure you are taking the readings properly.

2. Your teacher will set out five graduated cylinders containing different volumes of water. When it is time for your group to come up and read the five cylinders, record your measurements in the spaces below.

Data and Observations

CYLINDER #	1	2	3	4	5
Volume (ml)					

Ideas to Try

Try the ideas from **Beakers & Balances**.

© 1992 Rev 1999 The Wild Goose Co. WG 3003

Element Bingo

So What's Up?
Learn the elements of the periodic table by creating and playing the "Element Bingo" game.

"Why?" You Ask
Chemists use the element symbols as abbreviations for the chemical name. This saves a lot of time. Imagine having to write out every name for every element in all the compounds we use. Acetic acid, which is vinegar, has the molecular formula of CH_3COOH. Without the element symbols we would have to write Carbon-Hydrogen three times-Carbon-Oxygen-Oxygen-Hydrogen. A very slow method indeed. Element Bingo will help you memorize the common elements and their symbols. You can practice by matching one set of cards with the names of the elements to another set with the corresponding symbols.

Materials and Ingredients
2   blank bingo cards
1   sheet of colored paper
    scissors

The Procedure
1. On the first bingo card, fill out the names of the elements at random until you have all of your squares filled. Take one free square in the center. You can choose from any of the following elements:

H	Hydrogen	He	Helium	Li	Lithium
B	Boron	C	Carbon	N	Nitrogen
O	Oxygen	F	Fluorine	Ne	Neon
Na	Sodium	Mg	Magnesium	Al	Aluminum
Si	Silicon	P	Phosphorous	S	Sulfur
Cl	Chlorine	Ar	Argon	Ca	Calcium
Fe	Iron	Co	Cobalt	Ni	Nickel
Cu	Copper	Zn	Zinc	Br	Bromine
Kr	Krypton	Sr	Strontium	Sn	Tin
Au	Gold	Hg	Mercury	Pb	Lead
K	Potassium	Ag	Silver	I	Iodine

2. On the other bingo card fill in the symbols of the elements until all of the squares are filled and yes, you may take another free square. Choose from the same list.

3. Using the colored paper, cut out forty squares the size of a square on the bingo cards.

4. You're now ready to play bingo. Place the two bingo cards in front of you. When the teacher calls out a symbol or an element by name, place a colored square to cover that name on your bingo cards. The first person to get five in a row either vertically, horizontally, or diagonally wins. Good Luck!

Element Bingo

Ideas to Try

1. Have the kids choose an element and make a mobile to hang in the room. They can include any pertinent information on the element, such as pictures or samples. Creativity is the order of the day. You may also want to have them give oral presentations about the elements for a change of pace.

2. ChemSpell is an interesting game to play. The object is to make as many words or parts of words as you can out of the periodic table of elements in five minutes. Examples include HI (Hydrogen and Iodine), NeCK (Neon, Carbon, and Potassium), you get the idea. After five minutes, have the kids trade words and "translate" them back into the elements they came from. The kids usually love this game.

3. Element show and tell is another relevant activity The object is to have the kids bring in actual samples of the elements and display them. The metals most heavily represented are tin, gold, silver, titanium, copper, and nickel. You may want to have the kids observe helium, neon, and other gases in glass tubes if the high school physics or chemistry teacher will allow you to borrow them.

4. A more difficult version of Element Bingo can be played. When you call out an element name, require the kids to place a square on the appropriate symbol. Conversely, you call the symbol and they place a square on the name.

Element Bingo Grid

B	I	N	G	O

Marshmallow Molecules

Ahhh. Edible Water.

So What's Up?

Ever made mountains out of mole hills?   Try making molecules out of marshmallows!

"Why?" You Ask

Molecules are made up of combinations of **atoms**. Depending on how the atoms are combined, they can have entirely different effects. For example, two oxygen atoms and one carbon atom hooked together are called carbon dioxide, which is a by-product of our bodies. Plants use this gas to construct other molecules they need. However, if you have just one oxygen atom and one carbon atom, you get carbon monoxide, a poisonous gas. In this activity, you will construct ten different molecules that are fairly common.

"Why?" You Ask

Molecular shapes give scientists information about how the compounds might react and why they do what they do. Without this information, chemists would not be able to make the amazing advances they make each day.

Materials and Ingredients

100 mini colored marshmallows
  1 box of toothpicks
  1 brain
  1 pencil

The Procedure

1. A molecular formula tells you what kind of atoms are in the molecule and how many there are. If you don't know the atomic symbols, this would be a good time to stop and ask your teacher. Some of the symbols you're going to use are:

C	carbon	Cl	chlorine	Na	sodium
H	hydrogen	N	nitrogen	Cu	copper
O	oxygen	S	sulfur		

2. The formula for water is H_2O. H stands for hydrogen and the 2 right after the letter tells you there are two hydrogen atoms in the molecule. The O stands for oxygen. If there is no number immediately following the letter, it means there is only one of that kind of atom in the molecule. Another example would be NH_3. This is ammonia, which has one atom of nitrogen and three atoms of hydrogen. On the following pages, there are several formulas for you to dissect.

3. To construct the molecules, choose one color of marshmallow to represent one atom. All of the oxygen atoms must be the same color in any given molecule. Since you have a limited number of colors to choose from, the colors can change from molecule to molecule.

Marshmallow Molecules

Data and Observations

Draw pictures of the following molecules after you have constructed models from marshmallows. Be sure to identify the different atoms.

NaCl	NaOH	NO_2	$C_6H_{12}O_6$
CO_2	HCl	CO	H_2SO_4

Questions for the Anklebiters

Take the chemical formulas for these different molecules and fill in the blanks.

	Molecule	Symbol	Atomic Name	# of Atoms
1.	CO_2	C	Carbon	1
		O		2
2.	H_2		Hydrogen	2
3.	HC_1			
4.	CO			
5.	H_2SO_4			
6.			Sodium	1
			Oxygen	1
			Hydrogen	1
7.	NaCl			
8.	NO_2			
9.	$C_6H_{12}O_6$			
10.		C		1
			Hydrogen	4

11. Describe the difference between a molecule and an atom. _____

12. Which is larger, an atom or a molecule? Explain. _____

13. What do the little numbers next to the letters in the formula of a molecule represent?_____

Physical vs. Chemical Changes

So What's Up?
Experiment with physical and chemical changes by transforming various materials.

"Why?" You Ask
A physical change simply means that you have rearranged the shape of the object. It can be characterized as tearing, wadding, rolling, stretching, flattening, or otherwise physically changing the shape without altering the chemical properties. If you wad up a piece of paper, you create a physical change. If you burn that piece of paper, you cause a chemical change. You can identify a chemical change if there is a change in color, state, magnetism, or if heat is produced or lost.

Physical changes include:
1. Wadding up a piece of paper.
2. Chopping a piece of wood in half with an ax.
3. Tearing a piece of cloth.
4. Cutting a hamburger with a knife.
5. Biting into an apple.
6. Flattening a lump of clay with your fist.
7. Hitting a baseball with a bat.
8. Sitting on a balloon and popping it.
9. Blowing bubbles out of a soap solution.
10. Opening a piece of wrapped candy.

Chemical changes include:
1. Digesting your dinner.
2. Removing grease with soap.
3. Cooking an egg in a hot pan.
4. Milk that has gone sour.
5. Lighting the wood you chopped in half on fire.
6. Starting your car engine and burning gasoline.
7. Putting acid into your sink to dissolve hairballs.
8. Making cookie batter and baking it.
9. Taking an antacid tablet.
10. Exercising and using sugars and fats stored in your body.

Materials and Ingredients
1 balloon	limewater
1 block of wood	matches
1 clear plastic cup	salt
1 ice cube	
1 length of string, 12"	
1 nail	
1 straw	

Hmmm, an imminent physical change.

Physical vs. Chemical Changes

The Procedure

Perform the tasks below and identify the experiment as a physical or chemical change.

1. Take the nail and scratch the piece of wood. Physical or chemical change?

2. Open the book of matches and tear out one of them. Physical or chemical change?

3. Light the match (USE ADULT SUPERVISION). Physical or chemical change?

4. Take the balloon and blow it up. Physical or chemical change?

5. Place the ice cube on the table in front of you. It should have melted a little while you prepared for the lab. Place the end of the string on the ice cube and roll it around a bit to get it wet. Sprinkle a little salt on the string, count to five, and lift the ice cube off the table with the other end of the string. Physical or chemical change?

6. Fill the clear plastic cup one-third full with limewater. Note the color and insert a straw. Blow air from your lungs into the straw for about three minutes. Do not drink the water! Note the color — Physical or chemical change?

Data and Observations

Record your observations in the spaces provided.

Expt #	Physical Change	Chemical Change	How You Can Tell
1			
2			
3			
4			
5			
6			

Questions for the Anklebiters

1. List five physical changes you see in everyday life.
   a. _____
   b. _____
   c. _____
   d. _____
   e. _____

2. List five chemical changes you see in everyday life.
   a. _____
   b. _____
   c. _____
   d. _____
   e. _____

Ideas to Try

Find two physical and two chemical changes that are common in everyday life and either share them or demonstrate them for your class.

Chemical Reactions

So What's Up?
Determine if matter is created, destroyed, or just simply rearranged by taking a look at some chemical reactions.

"Why?" You Ask
Matter is never created or destroyed except in a nuclear reaction. Since these usually have government authorization, we are going to assume for our conversation the first part of that statement is always true. When a log is burned the wood turns to ash, water vapor, and several other gases. None of the atoms in the log are destroyed; they are just rearranged.

Materials and Ingredients
1 balance
1 balloon
1 test tube
  baking soda
  sand
  vinegar

The Procedure
1. Scoop a small amount of baking soda into the test tube.

2. Fill the balloon half-full with vinegar and attach it to the test tube. Be careful not to mix the two.

3. Place the test tube/balloon contraption on the balance and add enough sand to balance out the scale.

4. Mix the vinegar and baking soda. There should be a chemical reaction where gas is released and the balloon is inflated. Make sure the balloon stays attached to the test tube.

5. Weigh the mixture again and see if there is any change in the weight.

Data and Observations
Record your measurements in the spaces below. Be sure to include units.

Weight of contents before mixing	Weight of contents after mixing

Questions for the Anklebiters
1. Is the balance level after the reaction?

2. Is matter created or destroyed in this reaction?

3. Does matter weigh more, the same, or less after a chemical reaction?

 © 1992 Rev 1999 The Wild Goose Co. WG 3003

What's Inside?

So What's Up?
Use deductive reasoning and a toothpick to identify an unseen object's size and shape.

"Why?" You Ask
The depth the toothpick is inserted into the clay gives clues about the size and shape of the object. The firmness of the unseen object is determined when the toothpick touches it. Scientists often make decisions about the size and shape of objects without seeing them. You are using a scientific method called **deductive reasoning** to identify the unseen object.

Materials and Ingredients
  1 ball of clay with a small object inside
  1 toothpick

The Procedure
  1. Have someone secretly wrap the clay around a small object and mold the clay into a ball. The object must be firm enough so the pick does not break it.

  2. Poke the toothpick into the clay ball about fifteen times. Do not change the shape of the ball.

  3. Determine the size and shape of the object inside the clay.

  4. Guess what's inside the clay ball.

Data and Observations
The size and shape can be determined and, if it's a familiar object, it can be identified. List the bits of information that help determine the size, shape, weight, etc. of the object in the clay.

Questions for the Anklebiters
  1. What data did the toothpick give you about the object inside the clay?

  _____

  2. How many times did you have to poke the object with the toothpick?

  _____

  3. What could you tell about the firmness of the object?

  _____

  4. Did the toothpick give any information about the weight of the object? Why or why not?

  _____

Jr. Boom Academy

Coffee Filter Chromatography

So What's Up?
Separate different pigments of ink into their base colors.

"Why?" You Ask
Chromatography is a process that separates out a mixture of chemicals. In this case, the chemicals to be separated are the pigments in ink. A solvent (acetone) is used to carry the pigments up a porous paper. As the solvent climbs up the paper, the pigments that are the lightest and the most chemically compatible with the solvent move the the highest points on the paper, while the heaviest and least compatible are left near the bottom.

Materials and Ingredients
3   different colored water soluble pens (including black)
1   beaker
1   coffee filter
1   pencil
1   pie tin
    water

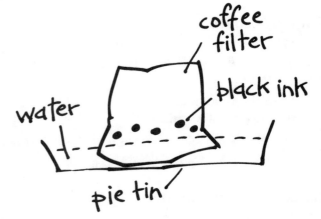

The Procedure
1. Open the coffee filter and with the pencil, draw a circle that is centered in the filter, but one inch inside the edge of the coffee filter. This is the line where you'll make the ink dots. Write the names (in pencil) of the colors you're going to use in the area of your ink spot placement.

2. Place ink spots on the areas you have identified.

3. Fill the bottom of the pie tin with just enough water to cover it an eighth of an inch deep.

4. Fold the coffee filter so it makes a cone and place it gently in the bottom of the pie tin. The bottom of the cone should be in the water, but the ink spots should be well above it.

5. Set the experiment aside and check it every five minutes to see which colors have separated.

Data and Observations
1. Fill in the data table below. In the time section, record the color at the top of the streak under each time indicated.

Original Color	Time (minutes)				
	5	10	15	20	25

Kitchen Chemistry

Floating Paper Clip

So What's Up?
Float a paper clip in a water-filled glass. What keeps it on top?

"Why?" You Ask
Water molecules are naturally attracted to each other. To understand this, you have to know what a water molecule looks like. All atoms and molecules are very small. You can't see them with your eyes, a magnifying glass or even a microscope. Scientists have to use very powerful microscopes to see the atoms. Check out the picture of the water molecule to the right. The little atoms on the top look like Mickey Mouse ears and have a positive charge. The big atom at the bottom has a negative charge. Because the positives and negatives attract one another, the water molecules all line up and hang onto each other. This is what we call the "skin" of the water. When a paper clip is placed on the surface of the water, its weight is evenly distributed across it and the water molecules hold it up.

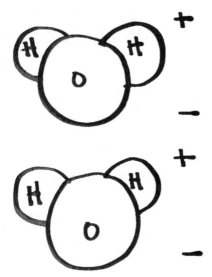

Materials and Ingredients
1   paper clip
1   pie tin
    water

The Procedure
1. Fill the pie tin as full as you can.

2. Carefully place the paper clip on the edge of the pie tin and gently push it out into the middle of the water. If you're careful, the paper clip will float on the surface of the water.

3. If you can't get the paper clip to float, place it on a section of toilet paper, then lower the paper into the water.

Questions for the Anklebiters
1. Why did the paper clip float rather than sink?

2. If the paper clip is bent out of shape, does it still float?

Ideas to Try
Try this experiment at home with other objects. May we suggest pins, upside-down tacks, pointy-side down tacks, hair pins, paneling nails, regular nails, pennies, and forks, for starters. You may not be able to get all of these items to float, but that's the nature of science. Failure is positive feedback.

How Full Is Full?

So What's Up?

Learn the meaning of "surface tension" when you stack as many water drops on the top of a penny as possible.

"Why?" You Ask

Water molecules have an attraction for one another due to their structure. This is described in the previous *Floating Paper Clip* experiment. Water also has properties called **cohesion** and **adhesion**, which allow it to cling to all sorts of things including other molecules of water.

Materials and Ingredients

1. cup
1. penny
1. pipette
   water

The Procedure

1. Fill the cup with water.

2. Place your penny on a flat, dry surface.

3. Dip the pipette in the water and fill it. Add drops one at a time to the penny until the water spills over on to the table. When the penny starts to look full, carefully add each drop one at a time.

Prediction

I think I will be able to stack ___ drops of water on top of the penny.

I was actually able to stack ___ drops of water on top of the penny.

Scared Pepper

So What's Up?
Sprinkle pepper on water, add a drop of soap to the center of the tin, and the pepper races to the perimeter.

"Why?" You Ask
Water molecules are so attracted to each other that, when they link up, they form an invisible "skin" which allows the pepper to float. The liquid soap breaks up the party, bursting the "skin" of the water when it connects to the water molecules. Think of it as a pin popping a balloon. The soap also forms a film of its own, spreading out across the tin. The pepper gets pushed along at the point where water meets soap.

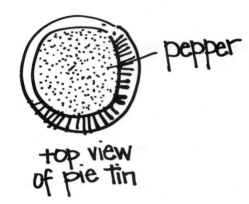

pepper

top view of pie tin

Materials and Ingredients

1   pie tin
    liquid soap
    pepper
    water

The Procedure
1. Fill the pie tin with water and sprinkle the pepper evenly over the surface.
2. Add a drop of liquid soap to the center of the tin. Draw a picture of what happens to the pepper in the pie tin below.

Questions for the Anklebiters
1. How did the soap cause all that movement?

2. If you put the soap at the edge of the tin, what happens?

Jumping Paper

So What's Up?

Fold a piece of paper into an accordion. Dampen the bottom fold and lower it toward the water. As the paper nears the water, watch what happens!

"Why?" You Ask

It's important to know what a water molecule looks like and how it behaves, to understand what is going on here. You'll find the explanation in the *Floating Paper Clip* experiment in this section.

Positives and negatives attract each other; water molecules tend to line up plus to minus. When the water molecules in the paper get close enough to the water molecules on the surface of the pie tin, they line up and attract, causing the paper to pull towards the surface of the water.

Materials and Ingredients

1 pie tin full of water
1 strip of white paper
   scissors

The Procedure

1. Cut a strip of paper from the edge of the sheet. It should be about one inch wide.

2. Fold the paper into an accordion shape, following the illustration to the right. Make the folds about an inch thick.

3. Lower the last fold to the water and get it wet. Lift the accordion out of the water and then lower it to the surface again but do not put it in the water. As you get closer and closer, you'll notice the paper will eventually jump into the pie tin.

Questions for the Anklebiters

1. How close does the paper need to be in order to jump toward the water in the pie tin?

2. Does this work if you use something other than water? Try milk, rubbing alcohol, corn syrup, or vinegar.

water

+ and ☐ attract

Milk Rainbow

So What's Up?

Fill a pie tin with warm, whole milk and add a couple of drops of food coloring in the shape of a triangle. Drop some liquid soap into the tin and watch the colors boil and swirl.

"Why?" You Ask

Milk is made up of water, vitamins, fats, and proteins. The last two of these ingredients are fairly sensitive to changes in the solution. If the solubilities are changed, the delicate bonds holding these molecules in place will be altered or destroyed. Adding the soap causes the milk to react with the soap and alter solubilities. This causes pandemonium with the fat and protein molecules and they start to bend, twist, contort, and basically swirl into all sorts of shapes. The food coloring simply acts as an indicator of the changes that are taking place.

top view of pie tin

Materials and Ingredients

1 pie tin
food coloring, 3 or 4 colors
liquid soap
milk

The Procedure

1. Fill the pie tin with warm, whole milk. If you don't have whole milk, 2% works fine. Just make sure it's at room temperature.

2. Add drops of food coloring using the drawing as a guide.

3. Put a drop of soap in the very middle of your design and watch what happens to the colors in the pie tin.

Questions for the Anklebiters

1. Will this same reaction work with water or soda pop?

2. What happens if you put the soap on the outer edge of the milk in the pie tin?

Ideas to Try

Try the same experiment and use cold milk. Compare the reactions you get using skim, 2%, and whole milk. Try cream or half-and-half and see if there is any difference.

Jr. Boom Academy — 32 —

Rainbow in a Bag

So What's Up?

Mix three primary colors into as many secondary, tertiary and quaternary colors as you possibly can. Then, identify as many colors as you can by name.

"Why?" You Ask

White light is composed of all colors. When white light hits an object, that object absorbs some colors and reflects others. The reflected light gives the object it's unique color. When you mix together different colors of glop, you get a combination of the absorption and reflection properties of the separate colors. Not surprisingly, this leads to new colors.

Materials and Ingredients

16 cups of water
 4 cups of cornstarch
1½ cups of sugar
 1 large sauce pan
 1 large wooden spoon
 1 resealable bag per student
    food coloring (red, blue, yellow)

The Procedure

1. Bring the water in the pan to a boil and toss in the cornstarch and sugar. Heat until it thickens, stirring occasionally.

2. Divide the mixture into thirds. Add food coloring to make one batch red, one yellow, and the last one blue.

3. Put one heaping teaspoonful of each color into the bags and seal them up.

4. Have the kids squeeze their bags to mix the colors and form new colors.

5. Color the circles below so they match the primary colors in each bag.

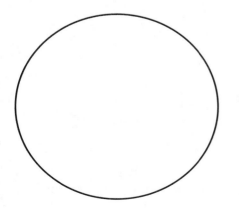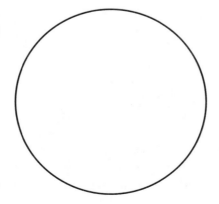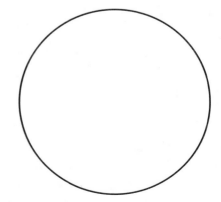

Jr. Boom Academy

Rainbow in a Bag

Mix the primary colors together, two at a time. Color the circles. These are the secondary colors.

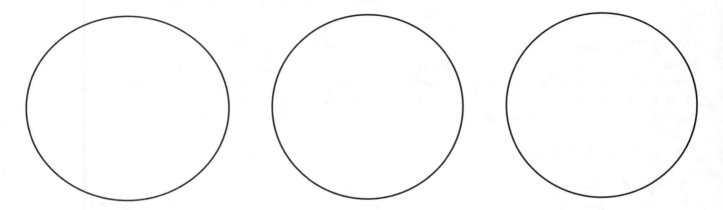

Mix the secondary colors together, two at a time. Color the circles. These are the tertiary colors.

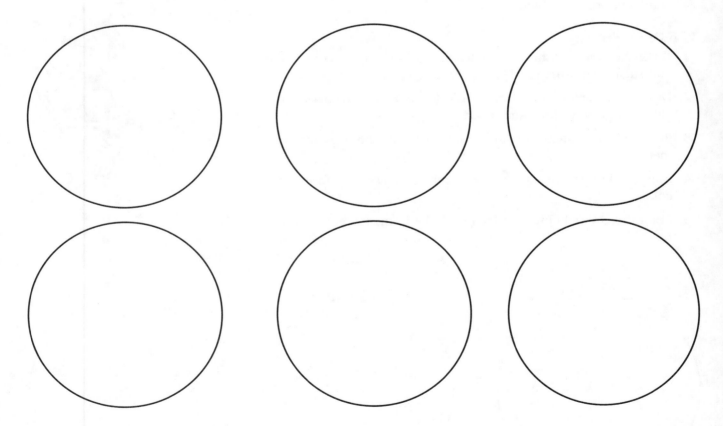

Ideas to Try

Instead of adding food coloring to the whole batch, add uncolored mixture directly to the bag first. Next, add drops of each shade of food coloring to the bags and have the kids work the food coloring into the mixture in order to get the desired mixed effect.

Vinegar Pennies

So What's Up?

Change the color of copper pennies by soaking them in vinegar (acetic acid).

"Why?" You Ask

Acid molecules in the air are a major contributor to air pollution. Cars, planes, ships, and other engines cause acid rain to fall, which erodes buildings and other structures. The effects of acid on copper roofing can be seen by using a bit of vinegar on a penny. When vinegar (acetic acid) reacts with copper it gives the copper a green coating. The acetate part of the acid combines with the copper on the pennies to form the green coating composed of copper acetate.

Materials and Ingredients

3-5 pennies
  1 paper towel
  1 saucer (or similar container)
    vinegar (acetic acid)

Watch Out For . . .

Acetic acid can cause irritation to skin, eyes, and nose. Use it carefully.

The Procedure

1. Fold the paper towel in half; fold it again to form a square.

2. Place the folded towel in the saucer.

3. Pour enough vinegar into the saucer to wet the towel.

4. Place the pennies on top of the wet paper towel.

5. Wait 24 hours.

Questions for the Anklebiters

1. Why does your mom's silverware get tarnished?

_____

_____

2. What would make the penny stay shiny even in the vinegar?

_____

_____

Ideas to Try

1. Pour out some vinegar and allow the water to evaporate for one day. Then compare the rate at which a penny turns green in the concentrated vinegar to the time it takes when you use vinegar from the bottle.

2. Try this reaction with other acids like hydrochloric or sulfuric.

Fishing for Ice Cubes

So What's Up?
Catch and lift an ice cube using a piece of string, some salt and not a single knot.

"Why?" You Ask
Salt lowers the temperature at which water freezes, so adding salt to ice causes it to melt. Sprinkling salt on top of the ice cube creates a layer of water, which is absorbed by the string. The ice *underneath* the string, not touched by salt, remains frozen. When the water absorbed by the string reaches the ice underneath the string, it has left much of its salt content behind and it re-freezes, freezing the string to the ice cube.

Materials and Ingredients
1 ice cube
1 length of string
1 pie tin
   salt

The Procedure
1. Rinse the ice cube under water for just a second and then lay it in the pie tin.

2. Place the string directly on top of the ice cube. Sprinkle a little bit of salt over the surface of the ice cube and the string.

3. Count to five and lift the loose end of the string and the ice cube with it.

Questions for the Anklebiters
1. If salt is used to melt ice in the winter, how come it caused the string to freeze to the ice cube?

2. Why do you mix salt with ice when you make homemade ice cream?

Hot Air Balloon

So What's Up?
Heat up some air molecules and watch your balloon expand!

"Why?" You Ask
Everything you see is made up of little particles called **molecules**. Even though you can't see these particles, they're constantly moving. They have tons of energy and are bouncing off one another all of the time. The hotter they are, the more energy they have, and the faster they move. As they move faster, they hit their surroundings harder. This often results in the molecules making more space for themselves.

Materials and Ingredients
1   balloon
1   candle
1   test tube
1   test tube clamp
    matches
    water

Watch Out For . . .
You must always use extreme care when working with fire. Absolutely no joking around and no horseplay!

The Procedure
1. Put one very small drop of water into the bottom of your test tube.

2. Put the balloon over the top of the test tube like the picture to the right. Make sure the tube is completely covered.

3. Have your teacher light your candle.

4. Put the test tube in the clamp. Heat the test tube over the flame and record what happens in the space provided.

5. Allow the test tube to cool down again. Did everything return to the starting conditions or did something change?

Hot Air Balloon

Data and Observations

Draw a picture of what the test tube looks like before and after it is heated.

Questions for the Anklebiters

1. Once you put the balloon over the test tube, is there any way for air to get into or out of the balloon?

_____

2. Did the number of atoms inside the test tube stay the same during the whole experiment or did it change?

_____

3. As you heat the test tube, are you adding energy to the atoms or taking it away?

_____

4. What happens to the balloon?

_____

Why?

_____

5. Are the atoms using more space or less when they are heated?

_____

Explain.

_____

_____

_____

_____

Starch in Food

So What's Up?

Discover how iodine can indicate the presence of another substance.

"Why?" You Ask

When starch comes into contact with iodine, the iodine, which is normally a reddish brown color, turns black. The black color is due to the formation of an I_3^- starch complex. Foods that are traditionally high in starch are potatoes, bread, any kind of pasta, and grains. Foods that are low in starch are meat and dairy products, fresh fruits, and some vegetables.

Materials and Ingredients

1 bottle of iodine
1 eyedropper
   paper towels
   several food samples

Watch Out For . . .

Iodine is a very dangerous chemical! Not only is it poisonous, but it stains everything it touches. DO NOT EAT any of the food samples used in this lab.

The Procedure

1. Put a small piece of food on a paper towel and put a drop of iodine on it. Record your observations. If the food turned black, there is starch present; if the iodine remained reddish, there's no starch.

2. Write the names of the food samples on your data table. Then, test each sample and record the results.

MMM, Yummers, black spaghetti.

Starch in Food

Data and Observations

Food	Color

If you're going to test more foods than there is space on the chart, create an additional data table and turn it in with your work.

Questions for the Anklebiters

1. List the foods from the bread group that tested positive for starch.

_____

2. List the foods from the meat group that tested positive for starch.

_____

3. List the foods from the vegetable and fruit group that tested positive for starch.

_____

4. List the foods from the dairy group that tested positive for starch.

_____

5. The health community has suggested that to live a healthier and longer life, you should eat foods that have lots of starch in them. Which food groups should you choose from?

_____

_____

Starch in Food

Ideas to Try

1. Have students develop a list of foods that are high in carbohydrates and starches and are also healthy. Also, have the students check labels and analyze the nutritional values of different foods.

2. Invite a dietitian to class to talk about the benefits of a healthy diet.

3. Have students keep records of their diets for a short period of time (2-3 days). When the time is up, discuss good nutritional habits as they relate to proper health. In fact, you could have a contest and reward the kid that exhibits the healthiest eating habits. What kind of reward? Not a candy bar!

Jr. Boom Academy

Upside Down & Up

So What's Up?
Lower the air pressure inside a test tube by oxidizing steel wool.

"Why?" You Ask
As uncoated steel is exposed to air, it rusts, or **oxidizes**. This experiment demonstrates the oxidation process actually reduces the air pressure inside a test tube. The iron in the steel wool reacts with the oxygen in the test tube to form ferric oxide, Fe_2O_3, or rust. Removing oxygen from the air inside the tube lowers the air pressure in the tube, causing the water to rise inside.

Materials and Ingredients
1   beaker, 400 ml
1   piece of steel wool
1   test tube
     water

The Procedure
1. Place the steel wool inside the test tube and douse it with water.
   Be sure to soak the steel wool well. Pour the remaining water out of the test tube.

2. Fill the beaker with 100 ml of water and set the test tube upside down in the beaker.

3. Record your observations.

Questions for the Anklebiters
1. Draw pictures of the set-up before and after the experiment is completed.

2. Describe what happens to the test tube.

3. Is oxygen taken out of the air or added to the air during oxidation?

4. Why do you think the water was pushed into the test tube?

The Great Heat Mystery

So What's Up

Mix two chemicals in a bag filled with phenol red and water. Feel the heat as the bag expands.

"Why?" You Ask.

When calcium chloride is mixed in water, it splits apart forming calcium and chloride. When this happens, heat is released. Once it's apart, each chemical is free to react with other chemicals like sodium bicarbonate.

The chloride reacts with the sodium in the sodium bicarbonate to form table salt and carbon dioxide, a gas. This gas causes the bag to swell.

Finally, the red liquid in the bag is an acid-base indicator called phenol red, but has a fancy chemical name called phenolsulfonephthalein. When the solution is basic, this indicator liquid is red; when it becomes acidic, it turns yellow. As the carbon dioxide is released it acidifies the water, which then causes the indicator to turn yellow. Better than autumn in New England.

Materials and Ingredients

1   bottle of calcium chloride
1   bottle of phenol red
1   resealable bag
1   small plastic cup
     measuring spoons
     sodium bicarbonate (baking soda)
     water

Watch Out For . . .

The only chemical you need to watch out for is calcium chloride. Handle it with care. It can be irritating to the skin, eyes, nose and lungs.

The Procedure

1. Add one-half cup of water to the resealable bag.

2. To this water add two capfuls (about 1 tablespoon) of phenol red.

3. Measure out one-half teaspoon of sodium bicarbonate and one-half teaspoon of calcium chloride into the plastic cup.

4. Open the bag, quickly add the two chemicals. Close and seal the bag. Tip the bag so the solution collects in the corner and feel the tip heat up. Observe the reaction!

CALCIUM CHLORIDE + WATER

HEAT

Fire Extinguisher

So What's Up?
Prove that carbon dioxide is heavier than air and acts as a fire extinguisher when "poured" over a flame.

"Why?" You Ask
Vinegar (acetic acid) and baking soda (sodium bicarbonate) react chemically to produce carbon dioxide. Carbon dioxide is heavier than air, so it remains in a beaker when you produce it and it pours like water. Pouring carbon dioxide over a flame extinguishes the flame because it displaces oxygen, which the flame needs in order to burn.

Materials and Ingredients
1 beaker, 1000 ml.
1 candle
   baking soda
   matches
   vinegar

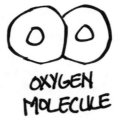

OXYGEN MOLECULE

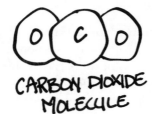

CARBON DIOXIDE MOLECULE

Watch Out For . . .
USE CAUTION WITH FIRE. Review the proper way to handle the matches and fire in general. Remember: Stop, drop, and roll.

The Procedure
1. Light the candle and place it where it can easily be seen by all.

2. Take a big scoop of air and announce that you're going to fill the beaker with air. Gently pour the air out on the candle and ask them to observe what happens. There should be no reaction, which is what you want.

3. Sprinkle the bottom of the beaker with baking soda. You don't need to be too precise during this experiment. This is what you would call "free-form" chemistry. You will get a feel for the reaction and adjust accordingly. Add a splash of vinegar. There will be lots of bubbles formed (this is the carbon dioxide being released), and they should come somewhere close to the top of the beaker. If you add too much vinegar, the bubbles will overflow onto the table. You now have a container full of carbon dioxide.

4. Hold the candle up and gently pour the gas from the container directly onto the flame. The gas will extinguish the fire. If you don't have success right away, it's because you're probably pouring over the flame. Back the beaker up a bit and try again. Pretend there is a stream of water flowing out of the beaker and you'll be able to visualize where to put the beaker.

Fire Extinguisher

Questions for the Anklebiters

1. What happened when the baking soda and vinegar were mixed?

   _____

   _____

2. Did the bubbles contain a solid, liquid, or gas?

   _____

   _____

3. Is the gas heavier than air or lighter than air? How could you tell?

   _____

   _____

4. What causes the candle to go out?

   _____

   _____

5. Is carbon dioxide a good fire extinguisher or a poor fire extinguisher?

   _____

   _____

6. What else could this gas be used for?

   _____

   _____

Ideas to Try
1. If you want to know the level of carbon dioxide remaining in the beaker, simply lower a match until it goes out. If you're careful, you can extinguish up to ten candles with one beaker of carbon dioxide.

2. Demonstrate a real fire extinguisher.

Toilet Paper Cannon

So What's Up?
Make a carbon dioxide cannon with baking soda, vinegar and a toilet paper tube. Watch the stopper fly.

"Why?" You Ask
The baking soda, or sodium bicarbonate, reacts with the vinegar, or acetic acid, to form carbon dioxide gas.

Materials and Ingredients
1   sheet of toilet paper
1   stopper
1   test tube
     baking soda
     vinegar

Watch Out For . . .
Errant stopper popping toward eyes.

The Procedure
1. Hold the toilet paper in your hand and sprinkle a pile of baking soda about the size of a quarter into the middle of the paper.

2. Fold the paper into a small package small enough to fit into the mouth of a test tube.

3. Fill your test tube one-third full with vinegar and then wiggle the toilet paper down into the top. Use the illustration below as a guide.

4. Stopper the whole mess and then give it a good shake, being careful to point it away from dogs, cats, and people.

Questions for the Anklebiters
1. Is baking soda a solid, liquid, or gas?

2. Is vinegar a solid, liquid, or gas?

3. What is produced when you see the two chemicals mix after shaking?

4. Is carbon dioxide a solid, liquid, or gas?

5. Why does the stopper shoot into the air?

Ideas to Try
1. Replace the stopper and see if it will shoot again.

2. Try other powders or liquids and see if the reaction can be repeated.

3. Experiment with the quantities of both vinegar and baking soda to determine the optimal "formula" for shooting the stopper the farthest.

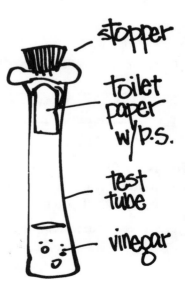

stopper

toilet paper w/b.s.

test tube

vinegar

Silly Spaghetti

So What's Up?
Make spaghetti dance in a glass of soda water.

"Why?" You Ask
Carbon dioxide is released from the solution and rises to the top of the container. If there happens to be something for the gas molecules to grab on the way up, they will. As the carbon dioxide accumulates on the surface of the spaghetti, it becomes more buoyant. When it gets enough gas molecules hanging onto it, it starts to rise toward the top of the container. When it gets to the top, the gas is released into the room and the spaghetti sinks to the bottom to pick up more gas molecules.

Materials and Ingredients
1   clear drinking glass
1   measuring tablespoon
     baking soda
     spaghetti, uncooked
     vinegar

The Procedure
1. Fill a large drinking glass three-quarters full with water and mix in a tablespoon of baking soda.

2. Add about two tablespoons of vinegar. This causes a lot of fizzing!

3. Break the uncooked spaghetti into small pieces and drop them into the glass. If the spaghetti doesn't rise and fall, add more vinegar until it does.

Questions for the Anklebiters
1. Why does the spaghetti rise to the top of the glass?

2. Would a piece of dry ice in water cause the same effect?

3. Why did the bubbles of carbon dioxide attach to the spaghetti?

4. Would air bubbles work as well as carbon dioxide?

Jr. Boom Academy

Crazy Raisins

So What's Up?
Toss raisins into a beaker of bubbling water and watch them rise and fall.

"Why?" You Ask
If an object immersed in liquid or gas is less dense than that liquid or gas, the object will rise. An example of this is a hot air balloon. When the balloon is filled with warm air, it is less dense than the surrounding air, so the surrounding air pushes the balloon up into the atmosphere. A bubble formed at the bottom of an aquarium will rise because it is less dense. A person in water who has his lungs full of air will also rise. When an object loses its buoyancy, it zips back down to the bottom of the container.

Materials and Ingredients
1 beaker, 400 ml
    vinegar
    baking soda
    raisins
    pencil
    water

The Procedure
1. Sprinkle enough baking soda just to cover the bottom of the beaker.

2. Add water to the beaker until it's four-fifths full. Stir the solution with your pencil until the baking soda is dissolved.

3. Place five raisins in the container and observe what happens to them.

4. Add enough vinegar to fill it completely. Observe the raisins.

Questions for the Anklebiters
1. Draw a picture of the beaker before and after the vinegar was added. Be sure to include the raisins.

Before   After

Crazy Raisins

Questions for the Anklebiters (continued)

2. What do you see forming on the raisins?

   _____

3. What two ingredients are causing the gas to form?

   _____

4. Why is some of the gas sticking to the raisins while the rest of it is bubbling to the top of the container?

   _____

5. Why are the raisins rising to the top of the container?

   _____

6. Why do they sink again after they get to the top?

   _____

7. Would this experiment work without vinegar? Baking soda? Water?

   _____

Ideas to Try

1. Test other items to see if they react the same way raisins do. If you'd like some ideas, try:

   a. balls of clay
   b. chunks of mothballs
   c. lima beans
   d. apple seeds
   e. pieces of broken spaghetti
   f. use your imagination

2. Use another solvent instead of water to dissolve the baking soda.

3. Construct the whole contraption in a very tall graduated cylinder. Any difference?

4. Gently tap the side of the beaker and see if you can get the bubbles to release the raisins before they get to the top.

Bubbleology

The Study of Bubbles

The Perfect Bubble Solution

Bubbleology is the scientific study of bubbles . . . no, not really. It's a word that's been made up for fun, but it does give you a good idea of what will happen in the next section. A soap bubble is simply a drip of water (with soap added, of course) that's been stretched out into a sphere.

Soap is an interesting molecule. It's actually a long chain of carbon, hydrogen, and oxygen atoms. At one end is a clump of atoms that like to be in the water (hydrophilic, or water loving), and at the other end is another clump of atoms that can't stand to be in the water (hydrophobic, or afraid of water). When soap is dropped into the water, the end of the molecule that likes to be in the water grabs onto water molecules, and the other end grabs onto soap molecules. As you've learned before, the soap makes its own film and this film holds soap bubbles together. Glycerin is added to make the walls of the bubble stronger.

A good formula to make your own bubbles is:

10 cups of water

1 cup dishwashing liquid (Joy® or Dawn®)*

1 teaspoon of glycerine

*According to full-time bubbleologists, these two brands of soap work the best when it comes to making long-lasting bubbles. The glycerine can be purchased at your drugstore, but it has been rumored corn syrup also works well. The longer you save your solution, the better it tends to get, kind of like fine wine. This section of the book is a little different in format as you will soon see. Have fun.

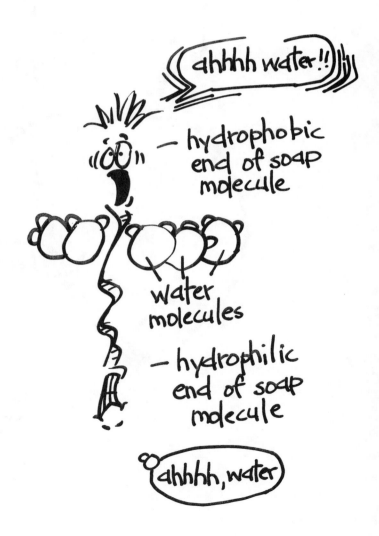

Floating Bubbles

So What's Up?
Float bubbles on a cushion of carbon dioxide.

"Why?" You Ask
The air we breathe is 20% oxygen and 80% nitrogen. We don't use the nitrogen; it goes in and comes out just the same. The oxygen is nabbed by the air sacks in our lungs and is transported to the bloodstream, where it is shipped out to different parts of our body, used, and traded for carbon dioxide. It is safe to say, however, most of the oxygen that enters our body leaves it unchanged (Big Clue). Dry ice is solid carbon dioxide. If left out in the open, the dry ice will disappear as a solid and exist only as a gas. Because carbon dioxide is heavier than air, an aquarium with dry ice in the bottom will hold carbon dioxide just as it will hold water. Soap bubbles are more dense than air, so they sink in air. But soap bubbles are less dense than carbon dioxide, so they float on the layer of carbon dioxide.

Materials and Ingredients
1 aquarium (empty)
1 bottle of bubble solution with wand
     dry ice, one or two pounds

Watch Out For . . .
The dry ice is extremely cold, on the order of 109°F below zero. If your skin comes into contact with a piece of dry ice for even a brief period of time, you run the risk of getting what is called a *cold burn*. The cells are frozen solid, killing them. If you must touch the dry ice, use a paper towel or gloves, if available.

The Procedure
1. Toss the dry ice in the bottom of the aquarium. If you take a hammer or other heavy object and smash it up a bit first, it will accelerate the reaction and release of carbon dioxide. Let the dry ice sublime (vaporize) for a couple of minutes.

2. Blow soap bubbles into the air away from the aquarium and observe where they go. Is it up, down, or do they stay right in the middle?

3. Once you have figured out that extremely tough question, blow soap bubbles over the aquarium so the bulk of the bubbles land in it. Do not blow into the aquarium or you won't get the desired effect. Record your observations.

Jr. Boom Academy

Floating Bubbles

Questions for the Anklebiters

1. First of all, draw a picture of what you observed in the box below.

2. Now describe what you just drew.

3. What gases are trapped inside the bubbles?

4. What gas is released into the aquarium?

5. Is the gas in the aquarium heavier or lighter than air? Is it lighter or heavier than the bubbles that land on top of it? How do you know?

6. Explain why the bubbles are floating in the middle of the aquarium.

7. Would this experiment work if the bubbles were filled with helium? Explain.

Ideas to Try

1. Slowly lower a match inside the aquarium and observe what happens. Why does the match go out?

2. Place a candle in the bottom of the aquarium and try to light it. Will this ever work? Explain, please.

3. Light the candle and place it on the table. Using an empty, dry beaker, scoop some of the carbon dioxide out of the bottom of the container and pour it onto the flame. Why does the candle go out?

4. Plop a couple of pieces of dry ice into a beaker full of water and observe what happens. Does this explain why the aquarium fills with carbon dioxide?

Six Pack of Bubbles

So What's Up?
Use a plastic, six-soda holder to make the best bubbles ever!

Materials and Ingredients
2   hands
1   plastic six-pack holder
1   tub of bubble solution

The Procedure
1. Grab the corners of the holder and dip them in the solution.
2. Lift and hold in the wind or make a large swoosh with your arms.

Ideas to Try
We started with a fairly basic idea for the first one. It gets trickier as you advance. The plastic rings are each coated with a thin film of soap. When the plastic holder is swished through the air, the movement creates bubbles. There are several different kinds of things you can use to achieve the same effect. If you have an old tennis or badminton racquet, dip that into the bubble solution and swish it through the air. Dip a chunk of cheesecloth and pull it tight as you swish it through the air. You'll see thousands of little bubbles. All you have to do is to check around the house and you will find lots of options.

Soup Can of Bubbles

So What's Up?
Make soap-bubble caterpillars by blowing bubbles through a soup can.

Materials and Ingredients
2-3 empty soup cans with both ends cut out
1 batch of The Perfect Bubble Solution
   breath control and practice
   packing or duct tape
   small shallow baking pan

The Procedure
1. Make up a batch of The Perfect Bubble Solution on page 52. Pour some into a shallow baking pan or similar container.

2. The soup cans need to be empty and have both ends cut out. Use the duct tape to join the separate soup cans into one, long, tunnel-like can. Make sure there are no sharp edges or protruding slivers of metal.

3. Dip the end of the cans into the solution and start to blow through the can. Ask the kids how far you should have your mouth from the opening. Before you continue, have the kids tell you why you should keep your mouth about six inches from the end, and review Bernoulli's Principle.

4. Blow a little, stop, blow a little, stop, etc. If you do this right, the soap film at the end of the can will seal itself each time you stop, and you can create a soap caterpillar.

Ideas to Try
A fun adaptation is to try cans of different sizes. Tuna fish cans are interesting, and the big industrial-size food cans called #10 tins yield monster bubbles. The one hang up is to make sure there are no sharp edges protruding or slivers of metal sticking out to pop the bubble as it starts to form. Empty toilet and paper towel tubes also work well until they get waterlogged and smoosh in on the ends. Basically, anything that makes a tube or opening will serve as a good bubblemaker.

Desk Top Bubbles

So What's Up?
Create bubble halves on a tabletop.

Materials and Ingredients
1   batch of The Perfect Bubble Solution (see page 52)
1   desk top
1   plastic straw
1   set of lungs

The Procedure

1. Pour a half-cup of bubble solution onto the desk top.

2. Put the straw in your mouth and hold it almost perpendicular to the surface of the table. Place the straw in the bubble solution.

3. Blow gently into the straw and a bubble will form on the desktop.

4. You can then remove the straw and start a second bubble next to the first. You can also start it inside the one you just blew. As the air is forced onto the table top, bubbles begin to develop. They increase in size until they pop. This happens because gravity gradually pulls the soap down towards the bottom of the bubble, which thins the top and sides until the bubble can no longer hang together.

Ideas to Try
Try making bubbles inside of bubbles. Blow a large bubble, gently remove the straw, and dip it into the solution on the table. Then reinsert it into the bubble. Begin blowing again and another bubble will appear. It is also possible to make bubble colonies, chains of bubbles, and two-story bubbles. Experiment and discover things on your own.

Electric Goomba

So What's Up?

Blow bubbles into the air while a partner uses a static charge to lead the bubbles around the room.

"Why?" You Ask

As the balloon is rubbed on the head or sweater of your partner, electrons are collected by the balloon. Electrons are teeny, tiny things that are part of atoms, and make up all the things in your world. Some of these electrons aren't held very tightly by their atoms, so they're easily stolen.

The water in the soap has both a positive and negative side, as explained earlier in the book. The balloon has a huge negative charge. Because the water molecules are fairly free to rotate, the negative charge on the balloon can attract the positive side of the water molecule and repel the negative side. The pushes and pulls between charges are stronger when the charges are closer. Therefore, the attraction between the balloon and the plus side of the water is stronger than the repulsion between the balloon and the minus side of the water. Overall, the soap bubble is attracted to the balloon.

Materials and Ingredients

2   kids
1   balloon
1   bottle of bubble solution with wand

The Procedure

1. Rub the balloon on the head of a lab partner or on a wool sweater in order to collect some random electrons.

2. Have a partner blow a nice bubble to take on a field trip.

3. Bring your charged balloon near the falling bubble. As soon as the bubble begins to move toward the balloon, move the balloon farther away so the bubble won't collide with it.

If you're not quick enough, the bubble will explode on the balloon. When this happens, the charge on the balloon is sometimes reduced or all together eliminated, and you'll need to recharge the balloon. If you have a hard time getting a charge on the balloon, it may be due to excessive humidity. Dry days work best. Also, if the person who's hair was rubbed used mousse or hair spray, it will inhibit the collection of electrons.

Questions for the Anklebiters

1. Why does the bubble get attracted to the balloon?

2. Can you devise a car or train that might use this means of propulsion?

Coat Hanger Bubbles

So What's Up?

Bend a coat hanger into a circle and make huge, hanging bubbles!

Materials and Ingredients

1   batch of The Perfect Bubble Solution
1   coat hanger
1   large baking pan or cookie sheet
1   pair of pliers
    masking tape

The Procedure

1   Pour the bubble solution into a large baking pan or cookie sheet.

2. The hanger forms the base for the solution. You will want to use a pair of pliers to get the hanger as round as possible. The problem point is usually where the hanger meets at the top and is twisted. Tape the sharp ends as much as possible to they won't pop your bubbles.

3. Dip the now rounded coat hanger into the bubble solution. Make sure you have a good film of soap on the hanger.

4. Now, as the hanger is pulled through the air, the soap film fills with air. With the twist of a wrist, the bubble is completed.

Questions for the Anklebiters

1. Do you think this would work with a lasso? How about a donut or bicycle tire?

2. What happens to the bubble on a windy day, a dry day, or a wet day? How about inside or outside a house?

3. Put your arm inside this bubble before you close it off? What happens?

String & Straw Bubbles

So What's Up?

Build a bubble-making contraption with two straws and a length of string. Send some beautiful big bubbles floating peacefully away.

Materials and Ingredients

2   plastic straws
1   length of string, 18"
1   tub with bubble solution

The Procedure

1. Thread the string through the straws and tie the ends of the string together.

2. Position the knot inside one of the straws and separate the straws so they look like the picture to the right.

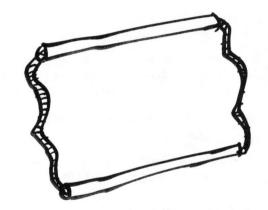

3. After you fill a tub with bubble solution, lower the straw bubble maker into the solution with the two straws held close together. A soap film will collect on both the strings and straws.

4. When the bubble contraption is removed and the straws separate, the soap film spreads out, forming a sheet. As the straws are pulled through the air, the air pushes on the soap film and forms the bubble.

5. To close the bubble, you must return the straws to their starting position so the bubble will be released from the magical mystery wand. If you don't do this, the stress applied to the bubble surface to seal it off is too great, causing the bubble to become unstable and burst.

Ideas to Try

You can also blow bubbles with this thing by creating a film or sheet of a bubble as described above and, while holding it in front of you, blow into the center of the sheet. A bubble will start to form. When it's as large as you would like, close the straws together and release the bubble.

Polymers, Colloids, & Random Synthetic Glop

Fried Peanuts

So What's Up?
Mix packing peanuts with acetone and watch them transform into a big blob.

"Why?" You Ask
Solvents are chemically compatible solutions of similar polarity to the stuff being dissolved (solute). They often have the same molecular composition and act to dissolve other compounds present. In this experiment, the acetone solvates the polystyrene, resulting in the collapse of the structure and the release of the trapped gas in the styrofoam that contributed to the colloid.

Translation: The acetone melts the plastic. Well, that's a bit simple-minded. The polystyrene is made up of lots of long molecules hooked together. When acetone touches these hooks, they fall off and the molecules fall together in a heap. As they fall, they squish the air out of the spaces, and all that is left is the plastic (polystyrene). Because there's a lot of air between the molecules, we can add a lot of peanuts to the beaker before it fills up.

Materials and Ingredients
1   bag of polystyrene peanuts
1   bottle of acetone
1   beaker, 100 ml
     latex or rubber gloves
     water

Watch Out For . . .
The acetone is very flammable and can cause irritation to the skin, eyes, and lungs. Handle carefully with adult supervision.

The Procedure
1. Put on your gloves.

2. Fill the beaker with enough acetone to cover the bottom about a quarter-inch deep.

3. Start adding peanuts and see how many you can get into the beaker before it is full. Be sure to keep track.

4. Pull the blob out and and rinse it under water to remove the acetone. Play with the blob and have fun!

Questions for the Anklebiters
1. Describe what "the blob" feels and looks like.

2. Why could you add so many peanuts to the beaker?

Baking Soda Foam

So What's Up?

Foam it up with a vinegar and baking soda, and liquid soap mix.

"Why?" You Ask

A **foam** is a colloid that results from a gas being dispersed in a liquid. In this case, CO_2 gas is formed by the reaction of acetic acid and sodium bicarbonate. This produces a **chemical foam**.

Definition time: Acetic acid is vinegar and sodium bicarbonate is baking soda. Chemists tend to like to know what it is that they are working with, while non-brain cramp types settle for household names. When you mix baking soda and vinegar, a gas called carbon dioxide is released. This is the same gas released when you pop the top off a soda. The liquid soap is what makes the bubbles in this experiment.

Here's the blow by blow. Liquid soap is added to the test tube and gets ready for action. When the vinegar is added, it finds its buddy baking soda and they high-five it. When they meet, they produce carbon dioxide, a gas, which is released in the solution. Lighter than the liquid, the gas will rise to the top. When it gets to the surface and tries to escape, the soap catches it and forms a foam. The gas is inside the liquid. We call this kind of colloid a chemical foam. Piece of cake. No, actually, tube of foam.

Materials and Ingredients

1   bottle of 10% sodium bicarbonate in water
1   paper towel
1   pie tin
1   test tube
   liquid soap
   vinegar

The Procedure

1. Fill the test tube half-full with sodium bicarbonate solution.

2. Add a dash of liquid soap to the test tube and gently shake the contents.

3. Quickly add vinegar to the test tube and hold it over the pie tin.

Questions for the Anklebiters

1. List three other foams that have seen around the house or school and what they're used for. (Psst, think of shaving, eating, and bathing).

2. How is a foam different from a solution?

Aluminum Foam

So What's Up?

Mix aluminum sulfate and baking soda for an expanding, sturdy foam.

"Why?" You Ask

A **foam** is a colloid that results from a gas being dispersed in a liquid. In this case, CO_2 gas is formed by the reaction of aluminum sulfate and sodium bicarbonate. Just like the last experiment, the sodium bicarbonate is actually plain old baking soda. Aluminum sulfate is used to grow some gnarly looking crystals for geology classes. The two chemicals react the same way as vinegar and baking soda. Together they produce a lot of gas and another chemical foam. This foam is a lot sturdier than the other foam you made — you'll see that when you flip the beaker upside down.

Materials and Ingredients

1   beaker, 100 ml
1   bottle of 14% aluminum sulfate, $Al_2(SO_4)_3$
1   bottle of 10% sodium bicarbonate in water
1   pipette
1   small pie tin
    liquid soap

Watch Out For . . .

The aluminum sulfate is acidic and may cause irritation to the skin and eyes. Always use with adult supervision.

The Procedure

1. Add a dash of liquid soap to the bottle of 14% aluminum sulfate solution. Swirl the solution to dissolve the soap.

2. Pour about one ounce of the sodium bicarbonate solution into the 100 ml beaker and swirl the beaker to mix the contents.

3. Place the 100 ml beaker in the center of the pie tin.

4. Quickly pour about one ounce of the aluminum sulfate/liquid soap solution into the 100 ml beaker.

5. Observe the reaction.

Question for the Anklebiters

How is this reaction similar to the foam produced from baking soda and vinegar?

Slush

So What's Up?
Make water disappear in an "absorbing" performance using dehydrated gel and water.

"Why?" You Ask
Several different kinds of gel are used as liquid absorbent in baby diapers and packing containers. If you check the Styrofoam™ packing containers of fine instruments like cameras and recording equipment, you will find little packages of water absorbent material. It is all the same thing.

Materials and Ingredients
3   opaque (non-see-through) drinking glasses
1   bottle of dehydrated gel (potassium polyacrylate)
    salt
    water

Watch Out For . . .
The potassium polyacrylate is not considered a harmful compound.

The Procedure
1. Prepare this experiment before your audience arrives. Place the three opaque glasses in front of you. Add just enough dehydrated gel to coat the bottom of the third glass with a thick layer.

2. Once your audience is in place, announce that you have three empty glasses. From an additional water source, pour about one-third cup of water into glass number one. Now, move the glasses around. Ask which glass contains the water. When they pick the correct glass, look inside and announce they are correct. Pour the water from glass one to glass two.

3. Repeat the mixing of the glasses and inquire again which glass has the water in it. Unless they're asleep, they will pick the correct one. Pour the water from glass two to glass three.

4. Repeat this one more time. When they guess the correct glass this time, hold it up and tell them they must be mistaken! Tip all three glasses over, one at a time being careful not to show the contents. The gel will have absorbed the water and will be stuck to the bottom of the glass.

5. To get the water out of the gel, add a generous amount of salt and stir. **Osmotic shock** will force the water out of the gel.

Ideas to Try
This stuff is great for retaining water in potting soil and can be used for germinating seeds.

Jr. Boom Academy

Pseudo Putty

So What's Up?
Create your own funny putty with white glue and borax.

"Why?" You Ask
A **polymer** is a chain of molecules all hooked together, like a tangled pile of jump ropes. The water molecules in the sodium tetraborate and the glue solution want to be buddies so they hook up. You start with jump ropes and end up with something kind of like a climbing net. Think of hanging with your good bud and being connected at the ears, shoulder, hip, knee, and your shoes are tied together. As soon as the two of you are connected, the water adds a third buddy and then a fourth, and so on . . . (this is all done very quickly). This chemical process gives you a compound that is rubbery. We call it **Pseudo Putty**; but people with brain cramps call it a polymetric compound of sodium tetraborate and lactated glue.

Materials and Ingredients
1 bottle of 50% white glue solution
1 bottle of 2% sodium tetraborate solution
1 plastic cup
1 craft stick
1 pipette
   food coloring

Watch Out For . . .
The sodium tetraborate may cause irritation. Always use protective gloves when handling chemicals.

The Procedure
1. Pour about one ounce of the glue solution into the plastic cup.

2. Add five drops of food coloring (optional) to the glue solution and stir well with the craft stick.

3. Draw one milliliter of sodium tetraborate into the pipette by filling it to the top mark. Then shoot it into the glue solution. Do this five times to add a total of five milliliters of sodium tetraborate. Stir vigorously with the craft stick for two minutes. Give it all you've got!

4. Remove the solid glob and roll it around in your hands to dry it off a little bit. It will remain sticky for one to two minutes but will eventually take on that elastic quality and transform into **Pseudo Putty**.

 © 1992 Rev 1999 The Wild Goose Co. WG 3003

Slime

So What's Up?
Mix two clear liquids together and slime appears out of nowhere!

"Why?" You Ask
Slime is a shimmery fluid polymer that is over 98% water. The water acts like a bridge, linking the polyvinyl alcohol to the sodium tetraborate through **hydrogen bonding**. The cross-linked polymer will shear if twisted and is endothermic as it flows, getting cold in your hands.

To simplify, there's these two molecules. They're floating around in some water and the water says, "Hey, you two should be buddies," and connects them to each other. They meet some other friends and they join together to make a chain of four. Then it gets wild with everyone chaining up with everyone else. Turns out they like each other so much that if you try to get them apart they don't want to separate and you have to literally tear them from each other. As for the endothermic part? That means it absorbs heat from your hand and becomes cold, due to the chemical change. Try moving it around — it will feel cool to your touch.

Materials and Ingredients
1   bottle of 2% polyvinyl alcohol
1   bottle of 2% sodium tetraborate
1   craft stick
1   plastic cup
1   pipette
    resealable bag

Watch Out For . . .
The sodium tetraborate and polyvinyl alcohol solutions may cause irritation. You should always handle chemicals with protective gloves.

The Procedure
1. Take the container with the polyvinyl alcohol and pour about one ounce of it into the plastic cup.

2. Fill the pipette with one ml of sodium tetraborate solution. Squirt the sodium tetraborate into the cup and stir like crazy with the craft stick. Suddenly you have slime.

3. Examine the properties of the cross-linked polymer (play with the stuff). If thicker slime is desired, repeat step 2 as many times as you want. (We like the slime made with two to three milliliters of sodium tetraborate).

4. When you're all done, you can save your slime by putting it in a resealable bag in the refrigerator. You can also just toss your slime in the garbage!

Solids, Liquids, and Gases

It's Not Out Till It's Out

So What's Up?
Observe the changes in candle wax from solid, to liquid, to gas with a fun trick.

"Why?" You Ask
This is a great demonstration to use for solid, liquid, and gas demonstrations, since all three phases of wax can be observed. When the candle flame is extinguished, the hot wax vapor rises. The vapor is extremely combustible and will burn if a lighted match is brought near it. When the vapor is reignited, the flame jumps down to the wick and the candle burns brightly once more.

Materials and Ingredients
1 candle
1 room without any air currents
matches

Watch Out For ...
Fire. If you're going to have the kids try this experiment, have them review the rules for dealing with fire before you begin. The experiment works best if there is little or no air movement; do your best to insure a stationary air mass in the room. Turn off all air conditioners and heaters, close windows and doors, and avoid excess breathing.

The Procedure
1. Light the candle and let it burn for a minute or two. This develops a nice pool of wax.

2. Strike a second match. Wet your fingers, quickly pinch out the candle flame, and look for the stream of smoke coming from the wick of the candle. Bring the lighted match so it comes in contact with the stream of smoke, about two inches above the wick. You'll see the flame jump from the match to the wick and reignite the candle. You don't want to blow the flame out or even lick your fingers and put it out that way, because you are reigniting the hot stream of wax vapor rising up from the candle. Blowing spreads out the vapor, and licking your fingers may dampen the wick too much. This vapor is highly combustible, and when the match is introduced into it, we have ignition.

3. The kids can observe the three phases of wax this way. The candle is **solid**, the **liquid** is just under the wick, and the **gas** phase is in the smoke that rises from the wick immediately after it is extinguished.

Questions for the Anklebiters
1. Is the wax in the candle a solid, liquid, or gas?
2. What happens to the wax that is heated just under the wick?
3. Is the smoke that rises from the wick immediately after it is extinguished flammable or inflammable?
4. Why do you think the candle relights even though the match never touches the wick?
5. Will this experiment work if there is no smoke rising from the wick? Why not?

Mothball Margaritas

So What's Up?

Heat up some naphthalene crystals in a test tube as you record ascending temperatures and note the results.

"Why?" You Ask

Some chemicals require a lot of energy to change states, and others just a little. Moth balls are used by people to prevent moths and other insects from eating holes in cotton and wool garments. The chemical name for moth balls is **naphthalene** and the melting point for this substance is very low.

The **melting point** of a substance can be detected by a flat spot in a temperature versus time curve as you heat the substance. As the chemical changes from one state to another, the temperature remains constant. Once the chemical has changed states, it then begins to show a temperature increase. If there are impurities in the substance, it may show more than one level spot as it is heated.

Materials and Ingredients

- 1 alcohol thermometer
- 1 beaker
- 1 candle, alcohol burner, or hot plate
- 1 test tube
  - matches
  - naphthalene crystals
  - water

The Procedure

1. Add several chunks of naphthalene crystals to the test tube. In fact, it should be about one-third full.

2. Place the thermometer inside the test tube. Begin heating the test tube and record the temperature every minute. After the naphthalene has fully liquefied, remove the thermometer and let the test tube cool down.

   This process will work best if the test tube is put into a beaker of water and the water is heated. A water bath will help maintain a more accurate temperature reading.

Questions for the Anklebiters

1. Record the temperature every minute until the naphthalene melts. Why did the naphthalene show a flat spot as it was being heated?

Time (min)	Temp °C	Time (min)	Temp °C
0		11	
1		12	
2		13	
3		14	
4		15	
5		16	
6		17	
7		18	
8		19	
9		20	
10		21	

Mothball Margaritas

Questions for the Anklebiters (continued.)

2. Graph the information

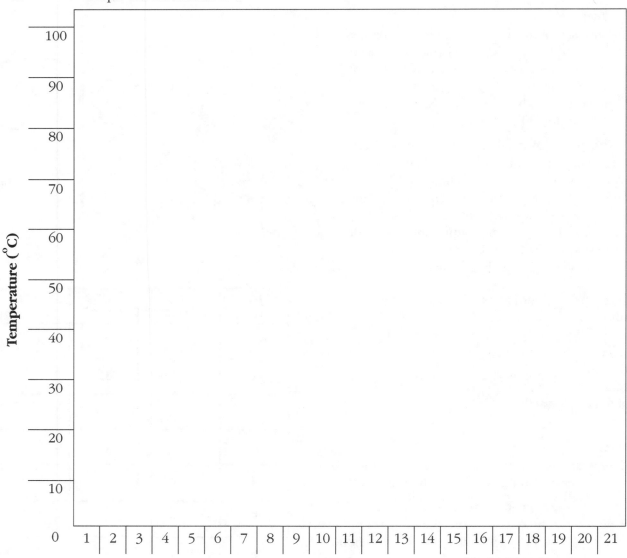

3. At what time did the naphthalene reach its melting point, and how could you tell?

4. Were there any impurities in the mixture?

5. What do you think would happen if you continued to heat the naphthalene until it boiled? What would the curve look like?

Steel Wool Sparkler

So What's Up?
A small piece of steel wool is teased out into a large, open ball. As a lighted match is brought near the steel wool, it catches on fire.

"Why?" You Ask
Iron can burn in the air if enough of the iron is exposed to heat and oxygen at the same time. When the iron is heated in the presence of oxygen, it oxidizes into ferric oxide. This process releases a great amount of heat and light, but it only works if the surface area is large enough.

Materials and Ingredients
- 1 pie tin
- 1 small piece of steel wool
- 1 test tube holder
-    matches

Watch Out For . . .
As always, use caution with fire.

The Procedure
1. Tease the piece of steel wool into a sphere about the size of a tennis ball. It should be as loose as possible. Clamp it into the test tube holder.

2. Light the steel wool on fire with a match. Hold it over the pie tin. The steel wool ignites because it has reached its **kindling point**, which is roughly 250 degrees Fahrenheit. The reason steel does not usually catch on fire is that it is packed so tightly the air is not available for combustion. By using the thin steel wool and teasing it out, you are providing enough air to the material for combustion.

3. The pie tin is there in case you feel the need to drop the test tube holder.

Questions for the Anklebiters
1. Why will the steel wool light and a thicker piece of steel, a nail for example, will not?

_____

2. What three things are necessary for a fire to start?

_____

3. Using this model, explain how a fuse might work.

Jr. Boom Academy

Sparkler on a Stick

So What's Up?
Discover the kindling point of iron filings. Didn't think iron would burn, did you? Think again.

"Why?" You Ask
The **kindling point** of a material is defined as the temperature at which that material bursts into flames. For regular old paper, that temperature is 451 degrees Fahrenheit. For steel, the temperature is only 256 degrees Fahrenheit. Everyone knows paper catches on fire before steel. So what's the deal? If steel ignites at a lower temperature than paper, why is it so hard to burn? Easy — there is more air mixed in with the paper than there usually is with the steel.

The fire is causing the iron particles to reach their kindling points. This is done because the particles are very small and exposed on all sides to a lot of air. As the particles heat and change to gas. They give off light and heat in the process. They also tend to propel themselves away from the stick due to the rapid expansion caused by heating.

Materials and Ingredients
1 bottle of iron filings
1 candle
1 darkened room
1 tongue depressor
  matches

Watch Out For. . .
The usual fire speech will do.

The Procedure
1. Light the candle and darken the room.

2. Dip the tongue depressor into the bottle of iron filings. This will coat the wood with lots of little goodies.

3. Roast the stick in the flame. Move it around so it doesn't stay in one place for too long and catch fire. You will be able to see lots of little sparkles shooting off the stick. It's quite entertaining, actually.

Questions for the Anklebiters
1. Why do small particles ignite, but when you try to light something like a pair of scissors or a nail, it won't burn?

2. Why do we see sparkles shooting off the stick?

3. What would happen to the stick if we held it in one spot in the flame too long?

Impatient Flask

So What's Up?

Confine liquid nitrogen in a stoppered flask and watch while it pops its top.

"Why?" You Ask

Liquid nitrogen (or oxygen) is obviously in a liquid state. Being exposed to room temperature and pressure, the liquid warms rapidly and changes into a gas. The more gas molecules you have in one spot, the greater the pressure. If this happens in a closed container, the pressure inside quickly increases to the point where any kind of stopper capping the flask shoots into the air.

Materials and Ingredients

1   flask, 400 or 600 ml
1   stopper to fit in the mouth of the flask
    liquid nitrogen

Watch Out For . . .

The liquid nitrogen is colder than you can imagine (-320°F below zero). Way cold. The hand that holds the flask should have a heavy glove to protect it from direct exposure just in case some of the liquid is spilled as you are pouring. Use nitrogen instead of oxygen, because anytime you have a concentration of oxygen and an opportunity for a spark, there is the possibility of combustion.

The Procedure

1. Pour the liquid nitrogen into the flask and stopper it.

2. Wait just a couple of seconds and the stopper will shoot into the air.

Questions for the Anklebiters

1. What causes the nitrogen to change states?

2. Why does the stopper shoot off the top of the flask?

3. How could you increase the rate of the reaction?

4. How could you decrease the rate of the reaction?

Ideas to Try

1. Take two identical flasks, add the same amount of liquid nitrogen to each one, and stopper them at the simultaneously. Put one in a pan of hot water and the other in a pan of ice water.

2. Experiment with containers of different sizes and diameters. Test tubes of varying sizes are always fun. Use a funnel to fill them. Rubber stoppers tend to work better than corks.

3. Set up a bucket or tub and figure out what kind of trajectory is needed to bomb it. You can also experiment with the amount of liquid nitrogen you put in the container. This will definitely affect the rate of the reaction.

Jr. Boom Academy

Ghost in a Manhole

So What's Up?
Warm up some air or liquid nitrogen molecules and make a coin do the jitterbug.

"Why?" You Ask
You can increase the pressure a gas exerts by increasing the number of gas molecules, or by making them move faster. Both things are going on in this activity. As the liquid nitrogen changes to nitrogen gas, you increase the number of gas molecules in the bottle. As these gas molecules warm to room temperature, they speed up. Thus, the gas inside hits the bottle and dime harder and more often, than before. Because the dime is pushed harder by the gas, it jumps up.

Materials and Ingredients
1 dime
1 bottle, one or two-liter
   liquid nitrogen or very cold air

Watch Out For . . .
The liquid nitrogen is extremely cold and will cause severe cold burns if it comes in to contact with your skin. Be very careful; this stuff is 320°F below zero and will freeze and kill skin cells instantly.

The Procedure
1. Fill a pop bottle with liquid nitrogen or, if you're low on that for the moment, put the pop bottle in the freezer for several hours.

2. Ask the kids to gently place the dime on the mouth of the bottle and observe what happens.

Questions for the Anklebiters
1. Are the contents of the bottle cold or warm?
2. Is the surrounding air warmer or cooler than the contents of the bottle?

3. What is happening to the contents of the bottle over a period of time?

4. Why do you put the dime on top of the bottle?

5. Why does the dime rattle and bounce once the lab has begun?

6. Will the dime do the same thing on a bottle that is at room temperature?

7. What would happen to the dime if it were placed on a bottle of extremely warm air and cooled rapidly by placing it in a pan of ice water?

8. How would the experiment be different if we placed a balloon over the mouth of the bottle instead of a dime?

cold air in bottle expands as it warms

Ideas to Try
1. Place a balloon over the mouth of the bottle and observe what happens as it is heated.

2. Place an inflated balloon over an empty bottle that has been heated in a pan of boiling water and is then placed in a bucket of cold water.

3. Set up a control experiment so the kids will have something to compare with the cool bottle.

San Francisco Summers

So What's Up?

Make instant fog with dry ice and a drop of water.

"Why?" You Ask

Mark Twain is quoted as saying one of the coldest winters he ever spent was a summer in San Francisco. If you have ever been there in the summer, there is a good chance you know what he was referring to. The Pacific Ocean never gets much warmer than 60°F in that part of the world, which means there is a cool air mass surrounding the city. As the summer warms up, the land is heated, which, in turn, heats the air over it. When warm air hits cool air, there is **condensation** or, in this case, fog.

Materials and Ingredients

1 beaker
1 candle
  dry ice
  matches
  water

Watch Out For . . .

The dry ice is extremely cold, on the order of 109°F below zero. If the skin comes in contact with a piece of dry ice for even a brief period of time, you run the risk of getting what is called a **cold burn**. The cells freeze solid and can't live. USE CAUTION WHEN HANDLING DRY ICE!

The Procedure

1. Fill the beaker half-full with water and drop a couple of chunks of dry ice into the water. The dry ice will **sublimate** (change directly from solid to gas) immediately. You are not actually seeing the carbon dioxide — it is fog. As the carbon dioxide is released from the dry ice, it is still very cold. As this very cold gas surfaces, it comes in contact with the warm air in the room, which also has moisture in it. That air cools rapidly, forming fog. As the fog spills out over the beaker and into the room, it warms up quickly and returns to the air as a gas.

2. Gently tip the container and allow some of the fog to fall out over the edge. You can pour this into an assistant's mouth.

3. Light the candle and pour some of the fog out onto it. The candle will be extinguished, just like the *Fire Extinguisher* activity.

Questions for the Anklebiters

1. Was it cold or hot gas escaping from the dry ice?

2. When this same gas reached the top of the water, was it cold or hot?

3. What happens when warm air meets cold air?

4. Why did the fog in the beaker appear just on top of the water and not under the water?

5. What caused the candle to be extinguished?

CO₂ Cannon

So What's Up?
Use dry ice to shoot a test tube "cannon."

"Why?" You Ask
Sublimation is the name given to a change of state when the material goes directly from solid to gas without bothering with the liquid phase. For our purposes, we're going to use **dry ice**, which is solid carbon dioxide. When dry ice is exposed to air at room temperature, it sublimes and produces tons of gas molecules. This causes the pressure inside the test tube to increase to the point where something has to give and the stopper is the easiest to remove.

Materials and Ingredients
1   test tube and stopper
    dry ice

Watch Out For . . .
The dry ice is extremely cold, on the order of 109°F below zero. If the skin comes into contact with a piece of dry ice for even a brief period of time, you run the risk of getting what is called a **cold burn**. The cells are frozen solid, killing them. Flying objects pose a minor threat to eyeballs so be careful where you aim.

The Procedure
1. Drop one to several small chunks of dry ice into the test tube.

2. Stopper the test tube with the stopper, rest it at an angle, and wait for the stopper to blow. Depending on the size of the container, how tight you place the stopper in the opening, the temperature of the room, and wind speed and direction, the stopper will shoot at varying distances and speeds.

Questions for the Anklebiters
1. Is carbon dioxide in the form of dry ice solid, liquid, or gas?

2. Is the air in the room warmer than the dry ice, or cooler?

3. Why does the stopper shoot off the test tube?

Ideas to Try
1. Put a little bit of very warm water in the test tube before you add the ice. Is the reaction accelerated or decelerated?

2. Put a test tube in a freezer for several hours. Pull it out to see if there has been an effect on the rate of the reaction. Why is it so much slower?

3. Will a test tube with a lot of dry ice shoot faster than a test tube with just a little dry ice? Why?

4. Try the same experiment with a regular piece of ice and compare the results.

Dry Ice Submarine

So What's Up?
Drive a "submarine" across water with carbon dioxide propellant.

"Why?" You Ask
As gas is released from a chunk of solid dry ice, the rapid increase in gas molecules increases the pressure. If there's a place for the gas to expand into, it will. If the direction of the expanding gas is controlled and directed, you can get it to work for you. This is what a rocket uses for thrust. The fuel is ignited, the hot gas expands, and if it is controlled, it can push the rocket up into the air.

Materials and Ingredients
1   stopper, one-holed
1   length of glass tubing with 45-degree bend
1   test tube
1   tub of water
    dry ice

Watch Out For ...
Dry ice burns at a temperature of 109°F below zero, so it kills skin cells rather quickly. Use protective gloves when handling the ice.

The Procedure
1. Push the tubing through the stopper as shown in the drawing.

2. Place the dry ice in the test tube and stopper it. Check the illustration above.

3. Place the test tube in the tub of water with the glass elbow joint in the water and let it go. Ooooooh aaaaaaah.

Questions for the Anklebiters
1. What is happening to the dry ice?

2. Why is there gas coming out of the piece of glass tube and shooting into the water?

3. What is causing the test tube to move forward?

4. How is this similar to a rocket?

5. What other applications can you think of for this kind of experiment?

Ideas to Try
Challenge the kids to invent cars or boats using gas as a propellant.

Grocery Bag Balance

So What's Up?
Experiment with different gases and determine which is heaviest.

"Why?" You Ask
Not all gases are created equal. Some catch on fire; others don't. Some give off beautiful colors when electricity is added to them and others don't. Some smell like rotten eggs — fortunately, most don't. Some are heavier than others. The weight of the gases depends upon the size and number of atoms in the gas molecule. Bigger and more means heavier.

Materials and Ingredients
2 paper grocery sacks
1 beaker, 1000 ml
1 bottle of vinegar
1 box of baking soda
    string
1 meter stick

The Procedure
1. Cut three 10-centimeter lengths of string.

2. Tie one to the center of the meter stick. Use the others to attach one grocery sack to each end of the meter stick. Hold the center string and adjust the two sacks so they're balanced. Use the illustration as a guide.

3. Sprinkle the baking soda so it just covers the bottom of the beaker. Add a splash of vinegar so the reaction fizzes up near the top of the beaker. Be very careful as you handle the beaker. It's full of carbon dioxide gas and any sudden movement will cause the gas to empty out of the beaker.

4. Have one person in your group hold the balance and another pour the carbon dioxide into one of the bags. It doesn't matter which one. Observe and record your observations.

Questions for the Anklebiters
1. Draw two cartoons below, showing what happened before and after the gas was added to the grocery sack.

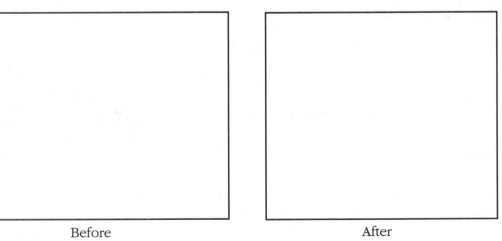

Before   After

Grocery Bag Balance

Questions for the Anklebiters (continued)

2. Is the baking soda a solid, liquid, or gas?

3. Is the vinegar a solid, liquid, or gas?

4. Is the carbon dioxide a solid, liquid, or gas?

5. The carbon dioxide is heavier than air. There are two ways you know this; describe each one.

Ideas to Try

1. Try the experiment again, substituting the following powders for baking soda, and describe your results:

   a. Flour _____

   _____

   b. Baking Powder _____

   _____

   c. Powdered Sugar _____

   _____

   d. Cornstarch _____

   _____

2. Try the experiment again, substituting the following liquids for vinegar, and describe your results:

   a. Water _____

   _____

   b. Lemon Juice _____

   _____

   c. Ammonia _____

   _____

   d. Vegetable Oil. _____

   _____

3. Open a two-liter bottle of soda pop and pour it into a large pitcher. Pour it as fast as you can until it looks like the foam is going to spill over the top of the container. Gently pour the gas from the pitcher into the grocery sack. Balance and explain your results.

Hidden Carbon Dioxide

So What's Up?

Unlock carbon dioxide from several different compounds with a weak acid.

"Why?" You Ask

Many compounds are classified as **carbonates**, or those that have the basic molecular component of carbon dioxide. The carbon dioxide can be released from these substances by mixing them with acids. Because carbonates react with acid to give off CO_2 gas, you can detect these compounds by the bubbles they contain.

Materials and Ingredients

6 test tubes

Samples of any or all of the following materials:

baking soda   marble chips
baking powder   vinegar or diluted sulfuric acid
chalk   washing soda
limestone

Watch Out For . . .

Acetic acid (vinegar) and the dilute sulfuric acid are both mild irritants to skin and eyes. If you get a little acid in your eye, flush with water for **15 minutes** and contact a physician. Wear your gloves.

The Procedure

1. Fill all six test tubes half-full with vinegar.

2. Add a scoop of baking soda, washing soda, and baking powder to each of three different test tubes. Record your observations each time.

3. Add small pieces of chalk, limestone, and marble chips to the other three test tubes. Record your observations.

Questions for the Anklebiters

1. Fill in the table below. If you did not test one of the things listed below, leave it blank.

Chemical	Reaction	CO$_2$ Present?
Baking Soda		
Washing Soda		
Baking Powder		
Marble Chips		
Chalk		
Limestone		

2. List the materials that have carbon dioxide present in them.

The Dip

So What's Up?

Cause a reaction between magnesium and sulfuric acid to form magnesium sulfate.

"Why?" You Ask

In the movie, *Who Framed Roger Rabbit*, the bad guy, The Judge, threatens the lives of cartoon characters who are otherwise immortal, with . . . The Dip. In the movie, The Judge used cleaning solvents. In this activity it's metal mixed with acid.

Materials and Ingredients

1   strip of magnesium ribbon, 1 inch in length
1   large test tube
    matches
    sulfuric acid, 0.5M
    test tube rack
    water

Watch Out For . . .

Acid is basically nasty stuff. If you get any on your skin, neutralize it with baking soda and soak under water for several minutes. If you get acid on your clothes, douse the area with water and plan on having a hole there. Also, never add water to acid! The water turns to vapor immediately under these circumstances and will expand to 1800 times its original volume.

The Procedure

1. Fill the test tube half-full with water and drop the magnesium ribbon into the tube.

2. Add another quarter-tube of acid. It should be three-fourths full at this point.

3. Touch the side of the tube and describe the temperature.

4. Put the test tube in a rack and bring a lit match close to the top.

5. When the experiment is finished, put the contents of the tube in a single beaker. Dilute the whole solution with a large quantity of water as you empty it into the drain.

Questions for the Anklebiters

1. What is forming on the magnesium ribbon?

2. What is responsible for the change in temperature?

3. What do you observe happening to the magnesium ribbon?

4. Why is there a pop when you bring a match near the top of the test tube?

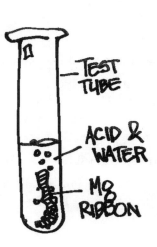

TEST TUBE

ACID & WATER

Mg RIBBON

Hindenburg Time

So What's Up?
Create hydrogen gas and zinc sulfate when mossy zinc reacts with sulfuric acid. Then, give it a very careful light.

"Why?" You Ask

The Hindenburg (a lighter-than-airship) was used for transportation before airplanes were more reliable. This huge balloon was filled with hydrogen gas. Hydrogen is the lightest atom on the periodic table of elements, so it floats extremely well. A gondola, which was hooked underneath the big balloon, held lots of people. The balloon would fly to different places. The only problem was that hydrogen is quite flammable. There are old movie clips that show the Hindenburg preparing to land. Just before it landed, it touched an electrical tower and exploded into flames.

Materials and Ingredients
    5   pieces of mossy zinc
    1   Ehrlenmeyer flask, 600 ml (with 300 ml water)
    1   beaker, 1000 ml (with 600 ml water)
    1   large test tube
    1   length of glass tubing, 6"
    1   length of rubber tubing, 18"
    1   rubber stopper, one-hole, for the flask
        glycerine
        matches
        sulfuric acid (0.5M)
        water

Watch Out For . . .
The acid is basically nasty stuff. If you get any on your skin, neutralize it with baking soda and soak in water for several minutes. If you get it on your clothes, douse the area with water and plan on having a hole there. No drinking this stuff!

The Procedure
1. Insert the glass tubing into the stopper before class, using a little glycerine as a lubricant.

2. Add the 300 ml of water to the Ehrlenmeyer flask.

3. Put the five pieces of mossy zinc into the Ehrlenmeyer flask. The water should cover them. Put your hand on the bottom of the flask and make a mental note of the temperature. Is it cool, warm, or neither?

4. Insert the rubber stopper into the flask.

5. Attach one end of the rubber tubing to the top of the glass tubing. Let the other end rest on the table.

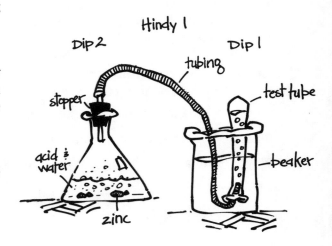

Hindenburg Time

5. Fill the test tube with water. Place your thumb over the top, tip the test tube upside-down, and lower it into the beaker full of water. When your thumb is under the water, let go. It's OK if a little air bubble is at the top of the test tube. Rest the test tube against the side of the beaker. It should be upside down and 99% full of water.

6. Place the other end of the rubber tube into the beaker and up into the test tube full of water.

7. You're ready for your teacher to add the acid at this point. Remove the stopper on the flask, raise your hand, be polite, say thank you, and experiment on, dudes and dudettes. Immediately after your teacher has added the acid, replace the stopper and feel the bottom of the flask.

8. You'll notice the hydrogen is collecting in the tube that is full of water. Hang onto the tube. As it fills with gas, it has a tendency to tip over.

9. Once the tube is full of gas, light a match and gently lift the test tube straight up out of the water. Tip the tube on its side, and at the same time hold the match at the mouth of the tube. Hindenburg time!

Questions for the Anklebiters

1. What happened to the temperature of the flask after the acid was added? What caused this change in temperature?

2. What did you observe happening to the mossy zinc after the acid was added? What chemical do you believe is responsible for this reaction?

3. Why was the gas filling the test tube? Why couldn't it push the gas out of the tube?

4. What happened when you touched the match to the end of the test tube full of hydrogen? What color is the hydrogen burning? If it's too light, darken the room.

Ideas to Try

1. Don't fill the test tube completely with water. Before you collect the gas, leave the top quarter filled with air. Increase it to half the tube and then three-quarters. What effect does oxygen have on the rate of combustion in the tube?

2. Try the same experiment, but use magnesium ribbon and hydrochloric acid. Try mossy zinc and hydrochloric acid. OK, now try magnesium ribbon and sulfuric acid. What conclusions can you draw from your observations about these four ingredients?

3. Remove the rubber tubing during the reaction and light the gas coming directly from the tube. Take a whiff of the gas coming out of the end of the tube. Smell like rotten eggs or your locker partner after he or she ate burritos for lunch? Thought so.

Miscibility of Alcohol

So What's Up?
Mix alcohol and water in different quantities and learn the meaning of "miscibility."

"Why?" You Ask
Miscibility is a term that describes two liquids that can be mixed in any ratio and not separate. When two or more compounds have similar polarities, they are often soluble in one another. This is true of water and alcohol. They are infinitely soluble in each other and are said to be miscible. The two combine in such a way the total mixed volume is less than the sum of the two separate volumes.

5 ml + 5 ml equals 9 ml?

Materials and Ingredients
6 test tubes
1 beaker of water
1 bottle of alcohol (isopropyl or ethyl)
1 graduated cylinder, 10 ml
1 test tube stopper
1 test tube rack
  sincere interest

The Procedure
1. Mix the following test tubes.

Test Tube #	Water (ml)	Alcohol (ml)	Total Mixed Volume (ml)
1	10	0	
2	8	2	
3	6	4	
4	4	6	
5	2	8	
6	0	10	

2. Stopper and shake each test tube one at a time. Set the test tubes in the racks and look at each combination of water and alcohol to see if any or all were miscible. Record your results in the space provided.

Questions for the Anklebiters
1. What evidence do you have to prove alcohol and water are miscible?

2. Experiment and find three different things that are not miscible with water.

The Rusty Balloon

So What's Up?
Reduce air pressure inside a test tube as steel wool oxidizes and takes the air right out of your balloon.

"Why?" You Ask
Oxidation is the process that allows oxygen to be chemically stolen from the air. This experiment demonstrates that the oxidation process actually reduces the air pressure inside a test tube.

As uncoated steel is exposed to air, it rusts, or oxidizes. The iron in the steel wool reacts with oxygen in the test tube to form ferric oxide, Fe^2O^3, or rust. Decreasing the number of gas molecules leaves the pressure inside the tube, so when the oxygen is taken out, low pressure created in the tube causes the balloon to shrink in size.

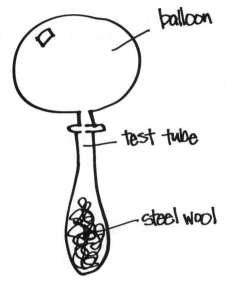

Materials and Ingredients
1 piece of steel wool
1 balloon, 9-inch
1 test tube
   water

The Procedure
1. Place the steel wool inside the test tube and douse it with water. Be sure to soak the steel wool completely. Pour the remaining water out of the test tube.

2. Place the inflated balloon over the test tube and set the experiment aside for a couple of hours.

3. Record your observations.

Questions for the Anklebiters
1. Draw pictures of the set-up before and after the experiment is completed.

2. Describe what happens to the balloon.

3. Is oxygen taken out of the air or added to the air during oxidation?

4. Why do you think the balloon was pushed into the test tube?

Jr. Boom Academy

Deep Freeze

So What's Up?

Show the crystalline nature of everyday objects with a liquid nitrogen "freeze."

"Why?" You Ask

Liquid nitrogen is about 320°F below zero. When an object comes into contact with the nitrogen, it invariably becomes **crystalline** in nature. If you were to drop a wine glass on a linoleum floor, it would shatter. The crystalline nature of the glass does not allow it to absorb much shock. When an object like a carnation, which has a great capacity for absorbing shock, is dropped on that floor, it tends to make out OK. If the temperature of that carnation is lowered to a point where the flower becomes almost entirely crystalline, and is dropped again, it will shatter like glass. Now you're getting the idea.

Materials and Ingredients

1   carnation or other object of destruction
    liquid nitrogen
    gloves
    large beaker

Watch Out For . . .

The liquid nitrogen is quite cold, as mentioned above. Use nitrogen instead of oxygen because anytime you have a concentration of oxygen and an opportunity for a spark, there is the possibility of combustion. If you do have an accident, do not hesitate to seek emergency medical attention. Liquid nitrogen is great for removing warts, but it can also do a lot of damage if used improperly. BE SURE TO WEAR GLOVES.

The Procedure

1. Demonstrate the elasticity of the carnation by hitting it on the table. It bends!

2. Place the carnation flower end down in a large beaker and hit it with a shot of liquid nitrogen. This should freeze the flower instantly. Wait an additional 60 seconds just to make sure every petal and leaf gets the deep freeze.

3. Pull the flower out by the stem and hold it up for temporary examination. Then whack it on the table. If everything proceeds according to Hoyle, the flower will shatter into little tiny pieces. Oooooh Aaaaaaah.

Questions for the Anklebiters

1. Why didn't the flower shatter when it was hit on the table the first time?

2. What happened to all the flower parts when the liquid nitrogen covered them?

3. What caused the flower to shatter? Was it the liquid nitrogen or the shock of the flower hitting the table?

4. Based on your observation, would you guess crystalline structures are elastic or non-elastic?

Ideas to Try

Favorite items to freeze :

a. hot dogs   b. racquetballs   c. oranges   d. rubber bands

Acids & Bases, pH Scale

The Litmus Test

So What's Up?
Determine whether a compound is acidic or basic using the litmus paper test.

"Why?" You Ask
Litmus paper changes color when it is introduced to acidic and basic solutions. Red litmus paper will change to blue when it comes in contact with a basic solution, and blue paper will turn red when it comes in contact with an acidic solution. Each of the papers will return to its original color when reintroduced to opposite solutions.

Materials and Ingredients
5 test tubes
2 vials of litmus paper (one red and one blue)
ammonia
lemon juice
washing soda (sodium carbonate)
water
white vinegar

Watch Out For . . .
Ammonia can cause nosebleeds and eye irritation. NEVER MIX AMMONIA AND BLEACH. It creates an extremely toxic nerve gas.

The Procedure
1. Fill each of the test tubes with a different chemical. The washing soda will have to be mixed with water and shaken until it dissolves.

2. Test each liquid with a strip of both red and blue litmus paper and see if it is an acid, a base, or neutral. Fill out the data table below.

Questions for the Anklebiters

Liquid	Red Paper	Blue Paper	Acid or Base
Lemon Juice			
Vinegar			
Water			
Washing Soda			
Ammonia			

1. Which of the liquids are acidic? _____

2. Which of the liquids are basic? _____

3. Which, if any, are neutral? _____

 © 1992 Rev 1999 The Wild Goose Co. WG 3003

Constructing a pH Scale

So What's Up?

Test several liquids for acidity with pH paper. Then, set up a pH scale.

"Why?" You Ask

pH paper is designed to respond to different chemicals' acidity. It identifies where chemicals are on the pH scale by using a series of colors. The scale ranges in color from deep red on the acid end to dark purple on the basic end. To give you a good idea of what the different colors look like, you'll test chemicals from one end to the other.

The acronym **pH** stands for per Hydrion, a sort of contraction for hydrogen ion. So a pH scale is a scale or measure of the available hydrogen ions in any given solution. That is a little deceiving because as soon as you cross over into the basic scale, you're really keeping track of the hydroxide ions. Lost? Don't worry. There are lots of juniors in high school all around the country who feel the same way you do right now. In either case, the severity of the reaction increases as you depart from the middle of the scale.

It's also important to know the scale is **exponential**, meaning the concentrations of ions increase by factors of 10. So something of pH3 is ten times more acidic than something of pH4. The most acidic reading is 1; something like sulfuric acid would be in the running. The most basic reading is 14 and those solutions are very basic. Right in the middle is a pH of 7, which is accepted as neutral.

Materials and Ingredients

1 pair of tweezers	ammonium hydroxide (ammonia)	lemon	saliva
1 pencil	apple	milk	sea water
1 sheet of paper	carrot	orange	tomato
1 vial of pH paper strips	grapes	plain soap	urine

The Procedure

1. Test each of the different compounds listed above. For the items like apples, grapes, and lemons, gently crush each of the different materials and test the juice. Be sure to use the tweezers each time you do this. Your fingers contain oils that will change the pH paper.

2. As you test each item, line it up on the sheet of paper in the order of the rainbow from red to purple. Be sure to write the name of each item you tested under the pieces of paper.

Constructing a pH Scale

Data and Observations

Record the pH of all the items you tested.

Item	pH	Item	pH
1		8	
1		9	
3		10	
4		11	
5		12	
6		13	
7		14	

Questions for the Anklebiters

1. Are fruits acidic or basic in nature?

2. What kinds of things would you think the pH scale could be used for?

3. Test the saliva in your mouth after you have eaten an orange and compare it with the color that was produced without eating the orange. Why the difference? Try a glass of milk.

Ideas to Try

1. There is an instrument called a pH meter. It's a handy little gizmo that will immediately tell you the pH of a solution. As with all good things, they range in price. It may be entirely possible to borrow one from the high school chemistry teacher.

2. Soil folks who deal with agriculture are always interested in the pH of the soil they're working with. Horticulturists and greenhouse managers worry about pH too. If there are any of these folks in your neck of the woods, tap the resource.

3. Have a gastroenterologist come to class and talk about acid stomachs. See if the local pharmacist will come out and talk about the body systems and the effects of drugs, legal and illegal, on the pH of the body.

Limewater Indicator

So What's Up?

Blow air into clear limewater and voilà! It changes it to something that looks like skim milk!

"Why?" You Ask

Calcium hydroxide is also called **limewater**. It is an indicator for carbon dioxide. When limewater is exposed to carbon dioxide, it turns from its usually clear, transparent color to a milky, skim milk color. The reason it becomes milky is because the product, calcium carbonate, is insoluble in water.

Materials and Ingredients

1 clear plastic cup
1 drinking straw
   limewater solution

Watch Out For . . .

Calcium hydroxide may cause irritation if it gets on your skin, so always wear your gloves. Also, don't drink the stuff. This may be obvious, but make sure you know how to blow, and not suck, with the straw. If you do accidentally drink some, contact a physician immediately.

The Procedure

1. Fill the plastic cup one-third full with limewater.

2. Insert the straw into the cup and blow gently into the straw. This creates bubbles in the limewater solution. Continue blowing until the limewater changes from clear to a milky color.

Questions for the Anklebiters

1. What color was the limewater when we started the experiment?

2. What color was the limewater at the end of the experiment?

3. What gas do we breathe into our lungs in order to survive?

4. What gas do we respire?

5. What do you think is responsible for the color change in the water, how do you know, and how would you check this?

6. Would you speculate this is a chemical change or a physical change?

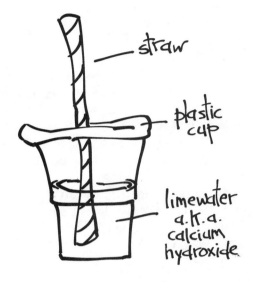

Make Up Your Mind

So What's Up?

Add sodium hydroxide (a basic solution) to methylene blue and dextrose and change the bright blue color to clear. Then, shake it up and see what happens!

"Why?" You Ask

The dextrose, or glucose, reacts with the basic sodium hydroxide to form a glucoside. The **glucoside** reduces the methylene blue dye to a colorless form. By shaking the tube, you're mixing oxygen into the solution, which oxidizes the colorless methylene blue to its original blue color. This reaction will change back and forth several times before the dextrose decomposes to a point of nonperformance.

Materials and Ingredients

1 bottle of 2% sodium hydroxide solution
1 bottle of methylene blue with 3.5% dextrose solution
1 pipette
1 test tube with cap
latex or rubber gloves

Watch Out For . . .

Sodium hydroxide may cause irritation if it gets on your skin or in your eyes. If you do happen to get it on you, flush with water immediately. If you accidentally drink the stuff, contact a physician immediately.

The Procedure

1. Slap those gloves on your paws.

2. Fill the test tube half-full with the methylene blue and dextrose solution. Slowly add one ml of sodium hydroxide. You'll notice the blue color disappears as the sodium hydroxide is added.

3. Replace the cap on the tube and shake it up and down vigorously several times. Observe the color change back to blue.

4. Let the test tube sit for a minute and the tube will clear. Shake it vigorously again and the blue will reappear. It just can't seem to make up its mind to be blue or clear. This reaction will continue for several hours until the dextrose gets tired and gives up catching oxygen molecules.

5. After a while, the intensity of the blue color starts to fade. Open the test tube so the solution can get some more oxygen and then cap it. Shake again and the blue color should deepen.

Questions for the Anklebiters

1. How long did it take for the blue color to disappear the first time?

2. Did the length of time get longer or shorter between color changes?

3. Why did you have to shake the tube to make it turn blue?

Red Cabbage Juice Indicator

So What's Up?
Steal the indicator out of ordinary red cabbage and turn it a different color with ammonia.

"Why?" You Ask
An **indicator** is a chemical that changes color when the pH of a solution goes up or down. Indicators are extremely large, complex molecules. When the pH of a solution goes up, hydrogen ions are lost and this changes the shapes of the molecule. When the shape of the molecule changes, the way it reflects light also changes. As the ammonia is added to the red cabbage juice, the pH goes up. It also changes the shape of the molecule producing the color.

Materials and Ingredients
2	glass test tubes	1	pipette
1	candle	1	test tube holder
1	kitchen knife		matches
1	leaf of red cabbage		water

Watch Out For ...
The ammonia may cause a lot of irritation if it gets on your skin or in your eyes. If you do happen to get it on you, wash it off. If you happen to get a snoot full, go get some fresh air and be grateful you don't live on Jupiter. If you accidentally drink the stuff, contact a physician immediately. Also, ask an adult to supervise when handling fire.

The Procedure
1. Chop the red cabbage into little pieces with a knife; the smaller the better because it will provide more surface area. Watch your fingers.

2. Place the cabbage bits in a test tube and add water until it becomes half-full.

3. Light the candle and hold the tube over the flame. Use the test tube holder to hang onto the tube so your fingers don't get toasty.

4. Heat the water and cabbage until the water turns a deep purplish red color. Remove the test tube from the flame and decant (pour) the liquid into a second test tube. Try to keep the cabbage bits out of the second test tube.

5. Take the cap off the ammonia and, using the pipette, add drops of ammonia to the red cabbage juice. Observe what happens to the color.

Questions for the Anklebiters
1. What color was the cabbage juice indicator? How about after adding the ammonia?

2. Why did you have to cook the cabbage to get the indicator?

3. What happens if you add some vinegar to the indicator? What about the ammonia?

Pink Petal Power

So What's Up?

Paint a white silk flower with a clear liquid. The flower turns bright pink. Cool!

" Why?" You Ask

The flower has been dipped in an indicator solution called **phenolphthalein** (pronounced fee • knowl • thay • lean). When this indicator comes in contact with basic solutions like sodium hydroxide, it turns a bright pink color. If you want to return the flower to its original white color, all you have to do is dip it in a weak acid like vinegar (acetic acid). This lowers the pH of the liquid absorbed by the silk flower and causes the phenolphthalein to return to a colorless state.

This reaction repeats itself quite easily. One word of advice: Let the flower dry completely between uses. This will give the chemicals a longer life.

Materials and Ingredients

1   bottle of 2% sodium hydroxide
1   small paint brush
1   white silk flower dipped in phenolphthalein
     latex or rubber gloves

Watch Out For . . .

Be sure to wear protective gloves. Sodium hydroxide may cause irritation if it gets on your skin or in your eyes. If you do happen to get it on you, flush with water immediately. If you accidentally drink the stuff, contact a physician immediately.

The Procedure

1. Dip the paint brush into the solution of sodium hydroxide and paint the petals of the silk flowers. Observe what happens when the liquid comes in contact with the fabric. Let the flower dry completely.

2. Try other household compounds with the permission of your parents. Experiment with things like ammonia, bleach, vinegar, cooking oil, Worcestershire sauce, window cleaner, and any other liquids you can find. CAUTION: NEVER MIX AMMONIA AND BLEACH. Make a list of liquids that are "basic" according to your flower test (those are the ones that turn the flower pink).

3. Dip your painted flower in a cup full of vinegar and see what happens to the pink color. Try cupfuls of water, ammonia, and soda pop, but remember, let the flower dry in between each dip. When you're done, save the flower for later.

Questions for the Anklebiters

1. What solutions caused the flower to become pink?

2. What solutions turned the flower white again?

3. How is the pH related to the color change?

Ideas to Try

Get some goldenrod paper and try the experiment with it.

Disappearing Ink

So What's Up?
Mix up some of that shifty, disappearing ink.

"Why?" You Ask
Thymolphthalein (pronounced thigh • mole • thay • lean) is an indicator which is used to determine the basicity of a solution. When it is added to a fairly strong basic solution (pH above 10.6), it is blue. When the solution is only moderately basic or is acidic (pH less than 9.0), it gets bashful and disappears. When the carbon dioxide in the air mixes with the solution, it lowers the basicity just enough to cause the color to turn tail and disappear. Since the blue color is created using sodium hydroxide that decomposes to sodium carbonate during the reaction, you have nothing to worry about if the ink gets on your clothes. After all, sodium carbonate is just washing soda, (a laundry additive). Ahhh, the wonders of science.

Materials and Ingredients
1 bottle of 2% sodium hydroxide
1 bottle of thymolphthalein in ethanol
1 paintbrush
1 pipette

1 white sock
   ammonia
   latex or rubber gloves
   white paper

Watch Out For . . .
Wear protective gloves. The ethanol solution is flammable and should not be handled around fire. It's also toxic, so don't drink it. Sodium hydroxide may cause irritation if it gets on your skin or in your eyes. If you do happen to get it on you, flush with water immediately. If you accidentally drink the stuff, contact a physician immediately.

The Procedure
1. Fill the pipette with sodium hydroxide. Add drops of sodium hydroxide to the bottle of thymolphthalein one at a time. After each drop, gently swirl the bottle to mix the two chemicals. When the solution turns dark blue, stop adding the sodium hydroxide. Too much of the chemical makes the ink very difficult to use.

2. Rinse the pipette thoroughly with water and fill it with the new solution you created (disappearing ink) in the thymolphthalein bottle. Squirt the disappearing ink on the white sock and observe what happens. Squirt some more ink in a different place; then blow on it to see if it speeds up or slows down the disappearing act.

3. Paint a secret message on a piece of white paper. When it has dried and disappeared, brush it with sodium hydroxide and see what happens. The message should reappear in the presence of any strong base such as ammonia. Throw the paper away when you're finished.

Questions for the Anklebiters
1. Why did the ink disappear and how long did it take?

2. At what pH value did the ink disappear?

Rainbow in a Bottle

So What's Up?
Add bromothymol blue to water for some interesting color switches.

"Why?" You Ask
Bromothymol blue, also known as dibromothymol sulfonephthalein, is an acid base indicator that operates in a pH range of 6.0 to 7.6. When it's in a basic solution, it's a blue color; when neutral, it's green; and when slightly acidic, it's yellow. As the acetic acid is added to the glass, it slowly acidifies the water. The more acetic acid that is added to the water, the lower the pH gets. When it reaches a neutral level, it turns green, and when it becomes acidic it turns yellow. When a couple of drops of sodium hydroxide (a base) is added, the solution becomes more basic, the pH rises, and the colors reverse themselves.

Materials and Ingredients
- 1 bottle of acetic acid (5%)
- 1 bottle of bromothymol blue
- 1 bottle of 2% sodium hydroxide
- 1 pipette
- 1 spoon (straight from the kitchen drawer)
- 1 tall, clear drinking glass
- latex or rubber gloves
- water

Watch Out For . . .
Put on your gloves. Sodium hydroxide may cause irritation if it gets on your skin or in your eyes. If you do happen to get it on you, flush with water immediately. If you accidentally drink the stuff, contact a physician immediately. The acetic acid is vinegar, but always use caution.

The Procedure
1. Fill the drinking glass with water and add one capful of bromothymol blue. Stir the water with a spoon to mix it thoroughly.

2. Using the pipette, add drops of acetic acid one at a time until a color change occurs. It will take a little patience but once it starts to change, it will turn quickly.

3. Rinse the pipette thoroughly in water. Slowly add drops of sodium hydroxide to the glass. Stir the contents of the glass every several drops and note the color changes. If you're careful as you add the base, you'll see the colors will exactly reverse. The solution will turn to green and then blue again.

Ideas to Try
Try using some pH paper to see if you can determine at which pH level each color change occurs.

Test Tube Mania

So What's Up?

Mix several chemical combos to produce dramatic color changes and lots of gas bubbles.

"Why?" You Ask

Each of the pH indicators; phenol red, crystal violet, congo red, bromocresol green, and neutral red; change color when acid or base is added. Their molecular shape changes to reflect or transmit different color light depending on whether they are in an acidic or basic solution. To identify the reaction with the reactant, match the following clues with the results of your tests and fill in the data table on the next page: Phenol red starts out as a red color and so does congo red, but congo red winds up turning blue, which is different than phenol red. Crystal violet starts out a violet color but is much different after the change. Bromocresol green definitely is not green either at the start or the end of the reaction, and neutral red can't decide if it should be orange or magenta. The licorice root solution contains baking soda, which is basic. The baking soda reacts with the acid to produce a lot of carbon dioxide gas and the licorice helps make a nice head of foam.

Materials and Ingredients

- 6 test tubes with caps
- 1 bottle of bromocresol green solution
- 1 bottle of congo red solution
- 1 bottle of crystal violet solution
- 1 bottle of licorice root with 10% baking soda solution
- 1 bottle of neutral red solution
- 1 bottle of phenol red solution
- 1 bottle of 2% sodium hydroxide
- 1 bottle of 0.5M sulfuric acid
- 1 pipette
- 1 test tube clamp
- 1 pie tin
  latex or rubber gloves

Watch Out For . . .

Be sure to wear your gloves. Sulfuric acid and sodium hydroxide may cause irritation if they get on your skin or in your eyes. If you do happen to get some on you, flush with water immediately. If you accidentally drink the stuff, contact a physician immediately. This acid is diluted so it is safe to use, but take care not to get it on your hands or clothes as it may irritate the former and put holes in the latter.

Test Tube Mania

The Procedure

1. Glove up! Locate the six test tubes. Fill them half-full with the solutions indicated in your materials list (bromocresol green, congo red, crystal violet, licorice root, neutral red and phenol red) and cap the test tubes. Remove the cap from the sulfuric acid and set the bottle in an area where it won't be knocked over.

2. To test each individual tube, remove the cap, clasp the tube in the test tube holder, hold it over the pie tin, and add sulfuric acid one drop at a time until you see a reaction. Be sure to note the beginning color and record the ending color of each tube.

3. The procedure for using licorice root (a brownish-yellowish liquid) is a bit different. To get the desired result, squirt a full dropper of acid quickly into the tube and observe the reaction.

4. When you're done testing and recording the data from all six tubes, go back and determine what was in each of the tubes. Use the clues in the second paragraph of the *So What's Up?* section.

5. Try reversing the reactions with a little bit of sodium hydroxide. If you're careful, you'll be able to reverse the reaction several times in a single test tube.

Data and Observations

Tube	Starting Color	Ending Color	Chemical Name of Indicator or Chemical
1	Yellow		
2	Yellow		
3	Red		
4	Red		
5	Violet		
6	Blue		

Questions for the Anklebiters

1. Why did the test tube with licorice root foam and not change color?

2. At what point do you know you've added enough acid or base?

Ideas to Try

Try using some ammonia, bleach, borax, or baking soda in the water along with the indicators to see if they change color. CAUTION: NEVER MIX AMMONIA AND BLEACH. As you experiment, keep track of the results in a data table.

Shake It Up, Baby

So What's Up?
Demonstrate "insolubility" with oil and water.

"Why?" You Ask
Remember the term **miscibility**? It's used to describe two solutions that mix well in any ratio (like alcohol and water). In this lab we're going to experiment with two solutions that don't mix very well. The term that applies to them is **insolubility**.

Water is a **polar** molecule. It has a positive and negative end. Because opposite charges attract, water molecules are naturally attracted to each other. If we look at a really oily molecule, we find it doesn't have charged ends. We call this a **non-polar** molecule. Water isn't attracted to oil, and won't mix with it.

Materials and Ingredients
1	beaker, 400 ml	1	test tube
1	graduated cylinder, 10 ml		cooking oil
1	pipette		liquid soap
1	rubber stopper		water
1	stirring rod		

The Procedure
1. Add 5 ml of water and 5 ml of oil to the test tube. Cap that baby and shake it for ten seconds. Examine the contents and record your observations below.

2. Shake the test tube for 30 seconds this time and observe the contents of the tube.

3. Add three drops of soap into the test tube (remove the stopper first) and then shake it for another ten seconds. Record your observations.

4. If your teacher decides you have enough materials, try stirring a beaker that has oil and water in it to see if that has any different effect.

Questions for the Anklebiters
Record your observations for each of the four test tubes.

1. _____

2. _____

3. _____

4. _____

Crystals & Minerals

Salt or Sugar Crystals?

So What's Up?
Mix salt and sugar together and using your hand lens, separate them from each other.

"Why?" You Ask
Sugar and salt look similar at first glance, but actually are quite different. Each has its own distinct shape and characteristics. Not only do they look different up close, but they melt at different temperatures, go into solution at different rates, and freeze at different temperatures.

Materials and Ingredients
1   hand lens
1   pencil
   salt (sodium chloride)
   sugar

The Procedure
1. Gently pour small amounts of each crystal onto the table and mix them together with your finger.

2. Using the magnifying glass and a pencil, separate the two kinds of crystals based on their shapes.

Questions for the Anklebiters
1. What difference in shape could you see between the salt and sugar crystals?

_____

_____

_____

_____

2. If you could not see the crystals, how else might you tell the difference between a bottle of sugar and a bottle of salt? What's another way?

_____

_____

_____

Ideas to Try
See if you can separate the two crystals based upon their solubilities in different solutions. We know both are soluble in water, but how about alcohol? You might also want to try things like acetone, vinegar, syrup, and soda pop.

Salt Crystals

So What's Up?
Use the process of evaporation to make some cool crystals.

"Why?" You Ask
Salt is one of the most common compounds on the earth. Its chemical name is **sodium chloride** and it is present in many of the foods we eat and plays an important role in keeping our bodies healthy. We are going to take a peek at how salt crystals form.

Materials and Ingredients
1   pie tin, small
    craft stick
    salt (sodium chloride)
    water

The Procedure
1. First, you're going to make a saturated salt solution by pouring some salt into the pie tin and adding enough water to almost fill it to the top. Stir the salt solution with the craft stick and watch it dissolve. If all of it disappears, add more salt and stir again. Keep this up until no more salt can dissolve. You now have a saturated solution.

2. Set the pie tin in the window and observe it over a few days.

3. In the box, draw a picture of the crystals that form in the tin.

Questions for the Anklebiters
1. What is the shape of the salt crystals?

2. Why did we make a saturated solution of salt?

Ideas to Try
Try doing the same procedure with sugar or alum.

Sweet Tooth Favorite

So What's Up?

Grow homemade sugar candy crystals in a jar.

"Why?" You Ask

A **supersaturated** solution is one that has as much solute (sugar) dissolved in the solvent (water) as possible. This can only be done at a high temperature. If you didn't heat the water, you'd end up with a saturated solution.

Materials and Ingredients

2	cups sugar (you may need more)	1	quart jar
1	stove	1	sauce pan
1	pencil	1	spoon
1	piece of string		food coloring (optional)

The Procedure

1. Add one cup of water to a sauce pan on the stove. Heat the water in the pan to boiling.

2. As the water starts to heat, add two cups of sugar, slowly stirring as you go. You should be able to add all of the sugar but if not, don't worry. Add a few drops of food coloring if you choose.

3. Pour the sugar water solution into the quart jar.

4. Tie the string to the pencil. Lower the string into the jar, resting the pencil across the top. The end of the string should be touching the bottom of the jar.

5. Wait several days. The longer you wait, the larger the crystals.

6. Remove and enjoy.

Questions for the Anklebiters

1. What is the shape of the sugar crystals?

2. Why did we make a supersaturated solution instead of just a saturated solution?

Ideas to Try

Try doing the same procedure using salt or alum.

pencil

jar

string

solution

Eggshell Geodes

So What's Up?

Use an eggshell to simulate a gas pocket found in certain lava flows. Wait a few days and watch your own geode grow. Beats waiting a few million years.

"Why?" You Ask

Suppose you have a bunch of lava and the lava is ooozing down a mountain, cooling and forming gas pockets. These gas pockets get filled with ground water containing lots of dissolved minerals. The ground water recedes and the pockets empty out. The pockets are then filled with ground water, emptied again, filled and then emptied, filled — you get the idea. Each time the ground water leaves the pocket, it forgets to take some of the minerals it brought in. When this happens, a geode starts to form. Over time, a variety of different minerals can be deposited in the pocket, providing a wide range of beautiful colors.

By definition, a **geode** is a deposit of mineral matter that has crystallized in a lava pocket caused by a gas bubble. The outside of most geodes is brown, but the inside can range from plain to quite beautiful. Geodes can be found all over the western United States, wherever there are lava flows.

Materials and Ingredients

1   bottle of copper sulfate crystals
1   clear plastic cup
1   craft stick
1   egg shell, clean
    boiling hot water

The Procedure

1. Make sure your eggshell is clean.

2. Add some copper sulfate crystals to one-quarter cup of boiling hot water and stir. Keep adding crystals until no more dissolve when you stir the water.

3. Fill your eggshell with the copper sulfate solution and set the shell aside in a place where it won't be disturbed.

4. Observe as the solution evaporates over the next few days. When the solution has completely evaporated, you will have a homemade geode. Cool!

Ideas to Try

Try making geodes using other compounds like nickel sulfate, alum, and potassium ferrocyanide.

BBQ Crystals

So What's Up?
Use the process of evaporation to help make crystals. This particular set of crystals will take awhile to grow and will continue to grow for several weeks if you take good care of them.

"Why?" You Ask
Barbecue charcoal is a very porous material. This means it allows things to travel through it. When you pour a cup of water on the soil in your garden, it flows through the soil. The soil is porous. When charcoal sits in a tin of liquid, it absorbs the liquid. If it's water or another solution, the water evaporates from the top of the charcoal but leaves behind the heavier molecules called **salts**.

Materials and Ingredients
1   chunk of charcoal
1   pie tin

1   recipe of charcoal crystal solution (see below)
      ammonia, 10 ml
      laundry bluing, 50 ml
      salt, 50 ml
      water, 100 ml

The Procedure
1. Mix up a batch of charcoal crystal solution. Stir well.

2. Place a piece of charcoal into the pie tin and pour the solution over it.

3. As the solution evaporates, crystals will form on top of the charcoal. If you want the crystals to continue to grow, make more of the solution and add it to the tin. Be careful not to add it to the top of the charcoal — you'll crush the crystals!

4. For colorful crystals, add food coloring to the charcoal before you begin.

Questions for the Anklebiters
1. What was the purpose of the ammonia in the charcoal crystal solution?

2. Why was a chunk of charcoal used and not just a rock?

Ideas to Try
See if you can find other things that would work to support the growth of the crystals, such as sponges.

Snowstorm in a Tube

So What's Up?

Create your own snowstorm with alcohol and potassium sulfate.

"Why?" You Ask

A **solvent** is a chemical that dissolves another chemical. When you put hot chocolate powder in water it dissolves. The water is the solvent. However, not all things dissolve in water. We're going to use this idea to create a snowstorm in a tube.

Isopropyl alcohol (another kind of solvent) is added to a 10% solution of potassium sulfate in a test tube. Potassium sulfate crystals begin to form at the top of the test tube and fall to the bottom like a snowstorm. The crystals form because the potassium sulfate says, "Hey, I don't like this guy. I'm outta here!" (Translation: potassium sulfate is not soluble in isopropyl alcohol.) The crystals fall because they are more dense than water.

Materials and Ingredients

1   bottle of isopropyl alcohol, 70%
1   bottle of 10% potassium sulfate
1   pipette
1   test tube

Watch Out For . . .

Isopropyl alcohol may cause irritation if it gets on your skin or in your eyes. If you do happen to get it on you, flush with water immediately. It is also a poison, so be very careful with this stuff. If someone does happen to accidentally drink it, contact a physician immediately.

The Procedure

1. Open the bottle of 70% isopropyl alcohol and the bottle of potassium sulfate.

2. Fill the test tube three-quarters full with the potassium sulfate. Fill the pipette with some isopropyl alcohol. Add the alcohol, one drop at a time, to the test tube containing the potassium sulfate solution.

3. Swirl the test tube around and observe any reaction that may form. If there's no reaction (snowstorm), add another couple of drops of alcohol and swirl again.

4. Add more isopropyl alcohol to create a really raging snowstorm.

Questions for the Anklebiters

1. Why does the potassium sulfate crystallize when the alcohol is added?

2. Describe the appearance of the solution in the test tube when the isopropyl alcohol is added to the potassium sulfate.

Caffeine Crystals

So What's Up?
Change caffeine from a gas to a solid and produce some amazing crystals.

"Why?" You Ask
Sublimation is the term chemists give to chemicals and compounds that are in a big hurry to change state. We're not talking about packing the kids and dog and moving from Montana to Texas. "State", as in solid, liquid, or gas, or as in the **three states of matter**. If you sublime, you change from a solid to a gas so fast you don't have time to be a liquid. Or, if you change from a gas to a solid so fast you do not have time to be a liquid. The liquid stage gets left out.

Materials and Ingredients
1   candle
1   fat test tube
1   skinny test tube
    caffeine crystals
    matches

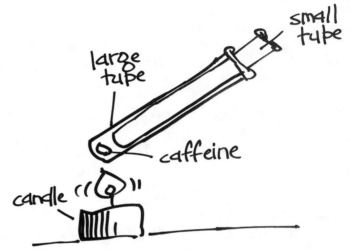

Watch Out For . . .
Caffeine may cause irritation if it gets on your skin or in your eyes. If you do happen to get it on you, flush with water immediately.

The Procedure
1. Dump the caffeine crystals into the big test tube.

2. Insert the small test tube into the big test tube so the closed end of the test tube is about an inch from the crystals.

3. Light the candle and hold the two test tubes over it. You'll notice the caffeine will start to melt and then turn into a gas. You don't have to worry about burning your fingers — the tubes never get that hot. However, it wouldn't be a bad idea to use good judgment.

4. Remove both test tubes from the flame and let them cool. As this happens, the caffeine vapor sublimates onto the bottom of the smaller test tube. Remove both tubes and examine the sublimed crystals of caffeine.

Questions for the Anklebiters
1. What is sublimation?

2. Why did the caffeine crystallize on the small test tube?

3. Are the sublimed crystals of caffeine purer than the unsublimed? Why?

Ideas to Try
Try using camphor crystals instead of caffeine.

Water Glass Garden

So What's Up?
Grow a crystal garden in a water glass.

"Why?" You Ask
As soon as the crystals enter the bottle, they are surrounded by the silicate. The silicate then reacts with the metal in the crystal, forming a new compound that begins to grow toward the top of the tube. The upward growth occurs because the pressure inside the bottle on the sides of the crystal is greater than the pressure on the top of the crystal. It has something to do with **semipermeable membranes** and **osmotic pressure**. If something is getting squished harder on the sides than on the top, it is going to grow upward. Makes sense, doesn't it.

The crystals produced are all metal silicates. In order of appearance and color, they are: copper (Cu) silicate which is blue, nickel (Ni) silicate which is green, and magnesium (Mg) silicate which is white. Use caution with these because in crystal form they could be very dangerous if you were to eat them. Bad idea.

Materials and Ingredients
3 vials with crystals of:
  $CuSO_4$ (blue)
  $MgSO_4$ (white)
  $NiSO_4$ (green)
1 bottle of sodium silicate
1 glass bottle with lid, 4-ounce

The Procedure
1. Open the bottle of sodium silicate and pour it into the glass bottle. Fill the glass bottle the rest of the way with water.

2. Open each of the small chemical vials one at a time and drop the crystals into the water/sodium silicate solution so the crystals fall to the bottom. Immediately replace the cap and discard the vial. These chemicals are very dangerous, especially if you have little kids around.

3. Observe the growth of the crystal garden. Make oooh aaah sounds as appropriate.

Questions for the Anklebiters
1. Why do the crystals grow upward?

2. Is there a limit as to how far the crystals will grow? What might make them grow farther?

3. Is this a physical or chemical change?

Heating of Cobalt Chloride

So What's Up?
Examine a compound that changes color depending on how many water molecules are attached.

"Why?" You Ask
Crystals of cobaltous chloride hexahydrate are naturally a rosy pink to red color. The word **hexahydrate** is easy to understand if you pick it apart. Hydrate stands for water molecules that are attached to a compound. So we know cobalt chloride has some water molecules hooked onto it, but we're not sure how many. It just so happens that the prefix **hexa** means six. So hexahydrate means six water molecules are hanging onto every one cobalt chloride molecule.

When you heat these crystals, they start to get warmer, and some of the water that is attached to them is released into the air. When this happens, the color of the crystal, which is red, changes to blue or lavender. Heat some more and more water is released, changing the color to violet. The blue-violet cobalt chloride is dissolved in acetone and, as the acetone evaporates, water from the air latches onto the cobalt chloride and turns it pink again.

Materials and Ingredients
1 bottle of acetone
1 bottle of cobaltous chloride hexahydrate
1 candle
1 pipette
1 pie tin, small
1 test tube
1 test tube clamp

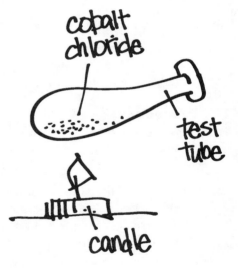

The Procedure
1. Light the candle and place the cobaltous chloride hexahydrate in the test tube.

2. Hold the test tube with the test tube clamp, and heat the cobaltous chloride over the flame.

3. Note the color change that occurs to the compound and any water that condenses on the sides of the test tube. When the chemical changes to violet, it is finished dehydrating.

4. After the reaction is complete, blow the candle out and allow the test tube to cool. Add a minimal amount of acetone to the test tube to dissolve the cobaltous chloride.

5. Pour the acetone solution into the pie tin and swirl the solution to aid in the evaporation of the acetone.

6. Note the color change that occurs when the crystals are formed.

Questions for the Anklebiters
1. Why do the crystals change color with heat?

2. Once the compound is blue and you leave it in the air, what will happen to the crystals?

3. Is this a physical change or chemical change? How do you know?

Minerals

So What's Up?

Demonstrate a mineral streak test and construct three common crystal shapes from paper patterns.

"Why?" You Ask

Minerals are the ingredients that make rocks what they are. If you think of a cookie as a rock, the ingredients that go into making a cookie can be compared to the minerals that go into making a rock. You have sugar, flour, chocolate chips, milk, etc. and when all of these things are combined and baked, you have a cookie. Rocks are the same way; minerals combine in different forms and quantities to make the rock. If you have different amounts and kinds of minerals, you have different rocks.

Minerals also have different properties. They can be identified by color, hardness, streak, crystal shape, and specific gravity, among others. There are over 300 different kinds of minerals and each mineral is different. In a sense, they produce their own fingerprint of characteristics. Some are very hard (like diamonds) and others are very soft (like gypsum). They come in every color and shape you can imagine. They also play a significant role in our world economy. Gold, silver, lead, copper, zinc, gypsum, and bauxite are all mined and used in metals or other products. Precious gems are prized for jewelry. We find diamonds, topaz, sapphires, rubies, emeralds, and other stones in necklaces, bracelets, and rings.

Materials and Ingredients

    5   mineral samples (azurite, malachite, gypsum, pyrite, galena, pectolite, and augite are possibilities)
    3   crystal patterns on paper (you have six to choose from)
    1   pair of scissors
    1   white porous tile
        clear tape

The Procedure

1. You will be examining several minerals. The first thing you'll do is record the color you see in the mineral.

2. The second thing is called the streak. This is determined by taking the mineral sample and drawing a line on a white, porous tile. The mineral will leave a colored line. Sometimes it's the same color as the mineral and other times it's completely different. Every mineral produces a specific streak color. Record this in the data box.

3. The third thing you're going to do is to construct three different crystal patterns characteristic of mineral shapes. The patterns will be provided by your teacher.

Minerals

Data and Observations

Record the following information about the minerals in front of you. There are no questions for this lab.

Mineral Name	Color	Streak
1		
2		
3		
4		
5		
6		
7		
8		

Draw pictures of the three crystals you have constructed once they are complete.

Ideas to Try

1. There are about ten different crystal patterns you will be able to find if you hunt around a bit.

2. Almost every major city and most out of the way towns out west have rock shops.   Invite the local rock hound shop owner out with a sampling of the goodies from their store.

3. Head on out to the local museum that has a rock collection.

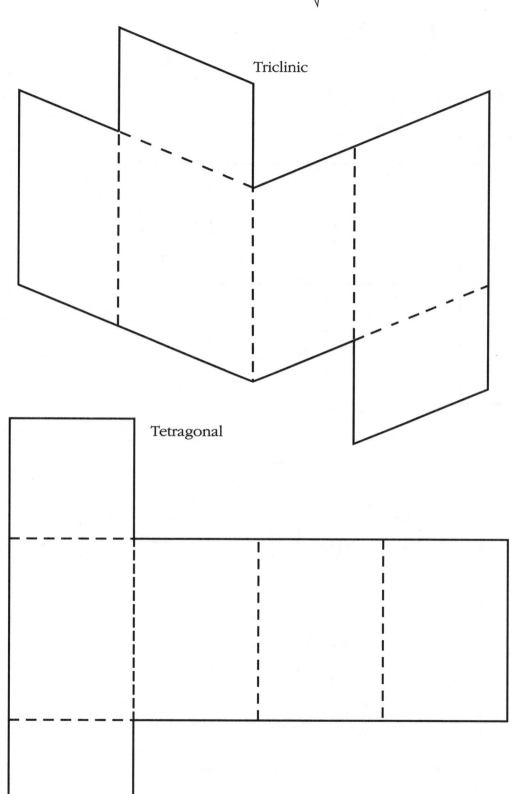

Triclinic

Tetragonal

Jr. Boom Academy

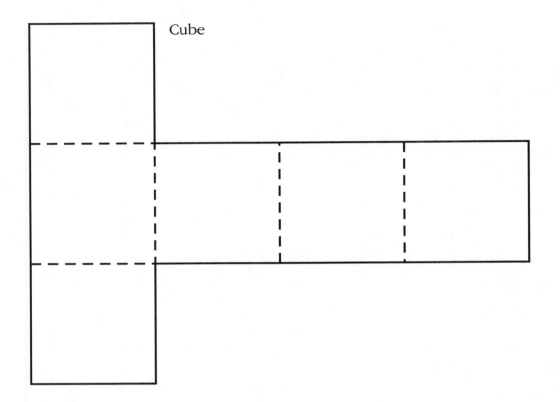

Cube

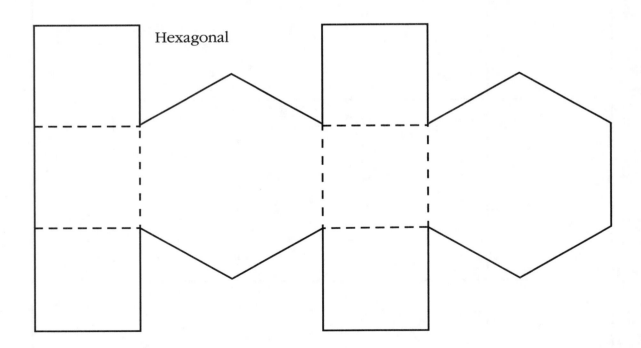

Hexagonal

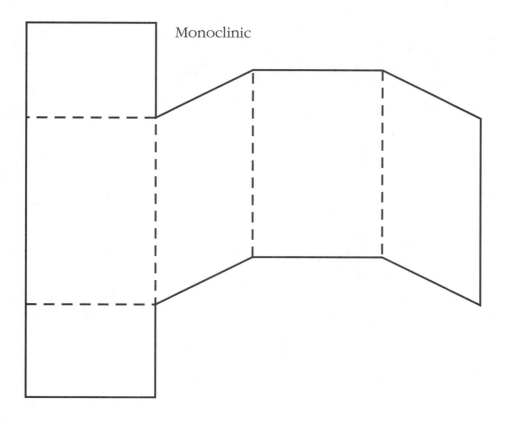

Monoclinic

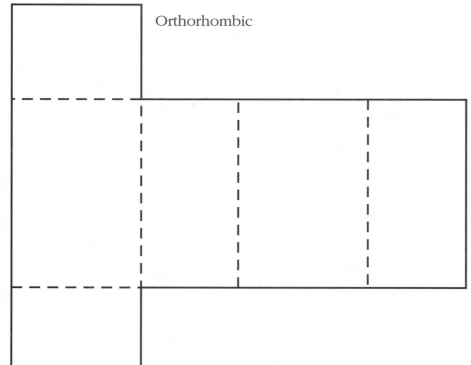

Orthorhombic

Oooh Aaah Chemistry

Conductivity

So What's Up?
Test the conductivity of several different solutions.

"Why?" You Ask
A **conductor** is a substance that allows an electric current to pass through it. Metal makes excellent conductors but plastic, wood, cloth, glass, and soap bubbles don't work very well (especially when you try to hook up test leads to the soap bubbles). Solutions are different critters. When many kinds of chemicals are in **solution**, they split apart and form ions. An **ion** is a charged particle. It can have either a positive or negative charge. When salt is in solution (mixed in water), the salt crystal dissolves and splits into a sodium ion, which has a positive charge, and a chloride ion, which has a negative charge. These charged particles have the ability to act as conductors.

Materials and Ingredients
 5   beakers, 250 ml
 3   alligator clips
 1   D-cell battery in battery holder
 1   light bulb in socket
 1   stirring rod
      baking soda
      diluted hydrochloric acid
      salt
      sugar
      vinegar
      water

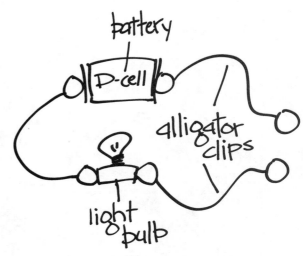

Watch Out For...
Both of the acids are diluted but will sting if you get them in a cut. They will also fade your clothing if you happen to spill it on them. If you do spill the acid, rinse the skin or clothing with water and if any irritation develops, contact a physician.

The Procedure
1. Hook the three alligator clips, the battery in the battery holder, and the light bulb together. Touch the two loose ends together and make sure the light bulb lights. This is a test to make sure your equipment is functioning.

2. To prepare the beakers, add 100 ml of dilute hydrochloric acid to beaker #1, 100 ml of vinegar to beaker #2, 100 ml of water to beaker #3 and add salt until no more will dissolve, 100 ml of water to beaker #4 and add sugar until no more will dissolve, 100 ml of water to beaker #5 and add baking soda until no more will dissolve.

3. Once all of the beakers have been prepared, you can test them to determine if they are conductors by placing the two test leads into the solution. If the light bulb comes on, then it means there is electricity flowing through the solution. If it doesn't, then you are out of luck. Be sure to whirl the test leads in the solution if the light does not come on right away. Record your results in the spaces provided.

Conductivity

Questions for the Anklebiters

Record your results in the spaces below. Tell if each solution is a conductor or nonconductor.

Solution #	Contents of Solution	Results
1		
2		
3		
4		
5		

1. What is a conductor?

   _____

2. How can you tell if a solution is a conductor?

   _____

3. Which solutions are conductors?

   _____

4. If a car battery must have a conductor inside it to charge properly, which solution or solutions would <u>not</u> be good to put inside a car battery?

   _____

5. Acids tend to be better conductors than bases. Lemon juice is acidic. Would it also conduct electricity? How could you test your idea?

   _____

6. The human body is a good conductor of electricity. Does this mean it possesses ions? Explain.

   _____

Electroplating

So What's Up??

Use the electroplating process to coat objects with a thin layer of metal.

"Why?" You Ask

All metal ions have a positive charge. When you run an electrical current through the solution that has metal ions, the ions migrate, or move toward the wires in the solution. In this experiment the copper ion is positive so it moves to the negative wire. Since the wire is attached to a paper clip, the paper clip gets copper-plated, or coated with copper ions. This is the same process used to gold-plate and silver-plate various products.

Materials and Ingredients

2	alligator clip jumper wires	1	nickel
1	bottle of copper sulfate	1	paper clip
1	clear glass half-filled with water	1	spoon
1	D-cell battery holder	1	straw
1	D-cell battery	1	toothpick

The Procedure

1. Place your D-cell battery in the holder and hook the two alligator clips to the battery terminal ends. Use the illustration as a guide.

2. Fill your glass half-full with water and add two capfuls of copper sulfate. Stir the solution with a spoon until all of the crystals have dissolved.

3. Hook the paper clip to the negative test lead (there is a minus sign on the battery) and place both the alligator clips in the tumbler solution. When the paper clip is electroplated, it will turn a bright copper color.

Questions for the Anklebiters

Test each of the items listed below and add some of your own if you like. Record the results of your experimentations in the space below.

Item	Yes	No
Toothpick		
Straw		
Nickel		

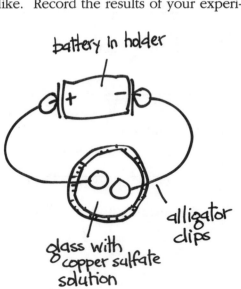

battery in holder

glass with copper sulfate solution

alligator clips

Electrolysis of Water

So What's Up?
Use electricity to split water molecules apart into hydrogen and oxygen.

"Why?" You Ask
Water molecules are not very big. They consist of one oxygen atom and two hydrogen atoms. If we could actually see them, they would look like an outline of Mickey Mouse's head. The atoms are held together by bonds, and the three atoms together are called a molecule of water. The molecule can be split up a couple of different ways. One, it can be split chemically. If water is added to another chemical, that chemical may use one of the hydrogen atoms for a reaction. Another way is that the atoms might get hit by a speeding particle and split into two or three pieces for a short period of time. The method that you're going to use today to split the molecule incorporates electricity.

Materials and Ingredients
2   alligator clips
2   test tubes
2   wood splints
1   battery, 9-volt,
1   battery clip, 9-volt
1   beaker, 1000 ml
    matches
    sodium sulfate, Na_2SO_4

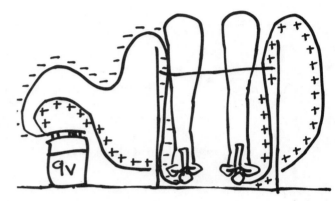

Watch Out For . . .
Fire. Always use caution when working with fire. Put the used matches in a safe, wet, place.

The Procedure

1. Hook the alligator clips to the battery, one alligator clip to each terminal.

2. Fill the beaker half-full with water and place the two alligator clips in the water.

3. Fill each of the test tubes with water. Place your thumb over the top, tip the test tube upside down, and lower it into the beaker full of water. When your thumb is under water, let go. It's OK if a little air bubble is at the top of the test tube. Rest the test tube against the side of the beaker. It should be upside down and 99% full of water. Repeat with the other test tube.

4. Place the end of each alligator clip up inside a test tube.

5. Add about one tablespoon of sodium sulfate and the reaction will proceed. As the test tubes fill with gas, be sure to compare the amounts of gas collected at each terminal.

6. Once the gas has been collected, test each of the gases for flammability. Light a match and use it to light a wood splint (it needs to be glowing — not burning), and gently lift the test tube straight up out of the water. Tip the tube on its side, and at the same time hold the splint at the mouth of the tube. If you hear a little pop, the gas is hydrogen. If the splint glows brighter, the gas is oxygen.

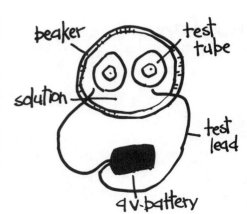

top view

Electrolysis of Water

Questions for the Anklebiters

1. What do you observe at each electrode once the alligator clips are hooked up to the battery?

   _____

   _____

2. Hydrogen ions are positively charged. Which electrode does it head for, positive or negative?

   _____

3. Oxygen ions are negatively charged. Which electrode does it head for, positive or negative?

   _____

   _____

4. What did you observe forming on each alligator clip?

   _____

5. Why did the tubes fill up with gas?

   _____

   _____

6. Which gas did you collect at the negative electrode and how do you know?

   _____

   _____

7. Which gas did you collect at the positive electrode and how do you know?

   _____

   _____

Ideas to Try

1. Try the experiment without the acid in the water.
2. Try a different kind of acid to see if all acids have the same effect.
3. Reverse the connections on the battery and see if the same gases are collected at the same poles.
4. Increase the amount of electricity by using a bigger battery and see what happens.
5. Find a way to mix the hydrogen and the oxygen together before you ignite them and see what happens.

Sun Ribbon

So What's Up?

Light a strip of magnesium ribbon and generate a super-bright, super-hot chemical reaction.

"Why?" You Ask

Magnesium is an element in the periodic table. It stores a tremendous amount of energy in its atoms, and, if its atoms are oxidized (burned) quickly, it produces a very hot, bright light. In fact, if you light a piece of magnesium ribbon in a darkened room, it will be like you are looking at the sun. As for hot, it burns at 5400 degrees Celsius.

Materials and Ingredients

  1 strip of magnesium ribbon, about 3 inches long
  1 test tube clamp
     matches

Watch Out For . . .

CAUTION WHEN USING MATCHES. You wouldn't want to touch the magnesium while it's burning. So much for the obvious! If the light produced by the burning strip hurts your eyes, be sure to look away.

The Procedure

1. Clip the magnesium ribbon into the test tube holder and darken the room.

2. Light the strip with the match. When you're all done there will be a white residue left on the holders. This is magnesium dioxide and can be wiped off without any problem.

Questions for the Anklebiters

1. Is there a lot of energy or just a little energy stored in the ribbon? How can you tell?

2. Why is there a white coating on the test tube clamp?

3. How could this reaction be used to help people?

Ideas to Try

The Dip is another activity in this book that uses magnesium ribbon.

Chemical Safety

Before conducting any experiment involving chemicals, you should have the proper equipment required. You should read and understand the chemical safety information contained in the Material Safety Data Sheets, a chemical safety text, and on the bottle labels of those chemicals you will be using.

In general, it's good practice to treat every chemical with proper handling methods and good lab safety practice. Always use protective gloves for your hands, goggles or a full face shield for your eyes and face, a lab coat or apron for your clothes and body, and a respirator or fume hood for those chemicals that should not be inhaled. You should also have eyewash bottles, fire extinguishers, emergency first aid kits, fire blankets, and chemical spill containment kits available.

The chemicals throughout this text are safe if used with reasonable care and with standard safety precautions. It's a good idea to know the physical, chemical, and toxicological properties of the chemicals you plan to use. Most chemicals are classified as irritants, no matter how safe they appear to be. Sugar, for instance, which has caused many a cavity and is great on cereal or in coffee, is a nuisance dust which may cause irritation to the eyes. It can also form combustible dust concentrations in air. However, we still recognize it as a safe chemical. The other classifications of chemicals are corrosive, flammable, and toxic. Toxic substances include carcinogens or poisons. Always use appropriate care and precautions.

For further information about health and safety, contact:

The Health and Safety Referral Service

American Chemical Society 1155 16th Street, NW

Washington, D.C. 20036

(202) 872-4515

Answers To Selected Questions

Marshmallow Molecules, page 19

	Molecule	Symbol	Atomic Name	# of Atoms
1.	CO_2	C	Carbon	1
		C	Oxygen	2
2.	H_2	H	Hydrogen	2
3.	HCl	H	Hydrogen	1
		Cl	Chlorine	1
4.	CO	C	Carbon	1
		C	Oxygen	1
5.	H_2SO_4	H	Hydrogen	2
		S	Sulfur	1
		O	Oxygen	4
6.		Na	Sodium	1
		O	Oxygen	1
		H	Hydrogen	1
7.	NaCl	Na	Sodium	1
		Cl	Chlorine	1
8.	NO_2	N	Nitrogen	1
		O	Oxygen	2
9.	$C_6H_{12}O_6$	C	Carbon	6
		H	Hydrogen	12
		O	Oxygen	6
10.		C	Carbon	1
		H	Hydrogen	4

11. *Atoms are the building blocks of molecules.*

12. *This is a bit of a trick question. The students will be tempted to say that molecules are larger than atoms because atoms are used to build molecules. It is possible, though, for an atom to be larger than a molecule. The smallest molecule is H_2, which is smaller than at least half of the atoms in the periodic table. It's the same as a very large brick being larger than an entire house made of very tiny bricks. Think Jack and the Beanstalk.*

13. *The numbers tell how many atoms of a particular element are in a molecule.*

What's Inside?, page 23

1. The toothpick can give data on where the object is, what its dimensions are, what shape it is, and whether it's hard or soft. It doesn't give any information about color, weight, or texture.

2. If the kids don't cheat and look at the object, they should have to poke at least 20 times to have a good idea of what the object is. If the object is difficult to guess, even 100 pokes might not be enough.

4. The toothpick doesn't give any information about the weight of the object, but a comparison of the weight of the clay with and without the object would do it.

Floating Paper clip, page 26

1. The paper clip floats because the water molecules underneath it are holding on to each other strongly enough to balance the force of gravity pulling the paper clip down. This happens only at the surface of the water (it's called surface tension), as you can see by dipping one end of the clip underwater (it sinks).

2. If you bend the clip out of shape, it will still float only if the new shape is still all in one plane. Any other shape tends to break the surface tension of the water.

Vinegar Pennies, page 33

1. It's unreasonable to expect the kids (or you!) will know exactly what's happening when silver tarnishes, but they might be able to figure out a chemical reaction is occurring. Silver tarnish looks somewhat similar to the copper acetate they just observed on the pennies. For the record, silver tarnish occurs when silver reacts with sulfur in the atmosphere (usually in the form of hydrogen sulfide gas) to produce silver sulfide.

2. The penny will stay shiny if you can somehow keep it from coming in contact with the vinegar. A coating of vaseline might work.

Fishing for Ice Cubes, page 34

1. Refer to the background information in the activity for an answer to this question.

2. Salt causes the ice around the future ice cream to melt, but what does that accomplish? To understand, you first have to realize the main goal in making ice cream is to remove heat from the liquid mixture as fast as possible. Having ice water around the container removes heat faster than just ice because there is more surface area to absorb heat when you have ice water. Ice water with salt in it is colder than ice water without salt, because salt lowers the freezing temperature of water. More surface area and colder ice water means the cream becomes ice cream faster.

Hot Air Balloon, page 36

1. Air can diffuse across the balloon membrane, but it's a slow process.

2. Ignoring the effect of diffusion, the number of molecules remains constant.

3. Heating the test tube adds energy to the molecules.

Upside Down and Up, page 40

3. During oxidation, oxygen is removed from the air and combines with a solid, in this case the steel wool.

4. Oxidation lowered the pressure inside the tube. The difference between the lower inside pressure and the higher outside air pressure then caused the water to be pushed up into the tube.

Silly Spaghetti, page 45

2. Putting a piece of dry ice in water would seem to cause the same effect, because dry ice is frozen carbon dioxide. However, the dry ice turns to gas so quickly it creates all sorts of bubbly commotion in the water, and you can't see the spaghetti rise and fall slowly. Once the dry ice is gone, so is the source of carbon dioxide.

3. Water naturally clings to some substances, such as spaghetti, in a process known as adhesion. The layer of water that surrounds each gas bubble adheres to the spaghetti, holding the bubble there. (Don't expect the kids to come up with this answer!)

4. Yes, air bubbles should work as well, providing you have a way of getting them to gradually release from a liquid.

Floating Bubbles, page 51

3. Air, composed mostly of nitrogen and oxygen, is inside the bubbles. If a person blows the bubbles, then there's an extra bit of carbon dioxide in there, too.

4. Carbon dioxide is in the aquarium.

5. The gas in the aquarium is heavier than air because it stays in the aquarium. The bubbles are less dense than this gas, because they float on top of it.

7. If the bubbles were filled with helium, which is less dense than air, they would float upwards towards the ceiling.

Baking Soda Foam, page 60

2. A foam is a gas dispersed in a liquid, and a solution is a solid dispersed in a liquid.

Mothball Margaritas, page 68

4. If the students' graphs show more than one level spot in the curve, that indicates the presence of impurities (things with melting points that are different from naphthalene).

5. At its boiling point, there would be another level spot in the curve. The temperature remains constant while the liquid changes to a gas.

Steel Wool Sparkler, page 69

1. A thick piece of steel won't light because not enough of its surface area is exposed to the air.

2. Fires require a fuel (steel, wood, paper, etc.), oxygen, and a catalyst (a heat source) to get started.

3. A fuse is something that stops the flow of electricity if the current gets too high. Metals get hot when electricity flows through them, so all you need is a piece of metal that will burn (thus breaking the electrical circuit) whenever the electrical current reaches a certain point.

Impatient Flask, page 71

1. It takes energy to change a substance from a liquid to a gas. In this case, that energy comes from the air in the room, which is considerably warmer than the liquid nitrogen and thus transfers heat to it.

3. You could increase the rate of reaction by adding more heat to the flask.

4. You could decrease the rate of reaction by cooling the flask.

Ghost in a Manhole, page 72

6. The dime won't do the same thing on a bottle at room temperature (with no liquid nitrogen inside). In order for the dime to dance, you have to increase the pressure inside the bottle. If the bottle is already at room temperature, it won't heat up and the pressure won't increase.

7. In this case, the pressure inside the bottle would decrease. All that would happen is the outside air pressure would tend to push the dime down towards the lower pressure inside the bottle. Since the dime is solid, nothing would happen except that maybe the dime would be hard to remove from the top of the bottle.

8. If you replace the dime with a balloon, the balloon will expand as the pressure inside the bottle increases.

The Dip, page 79

1. Magnesium sulfate.

2. The chemical reaction that takes place is exothermic, which means it gives off heat.

4. The pop is due to the explosion of hydrogen gas.

Hindenburg Time, page 81

1. The chemical reaction between the acid and the zinc is exothermic, meaning it gives off heat.

3. The gas is less dense than water, so water pushes it up to the top of the tube.

Limewater Indicator, page 89

3. Oxygen.

4. Carbon dioxide.

5. Carbon dioxide is responsible, but that's not obvious because our breath contains other gases. To check, try bubbling pure carbon dioxide into limewater.

6. This is a chemical change, because a new substance is produced.

Disappearing Ink, page 93

2. This would be difficult to determine experimentally, but you can do it using a pH meter. Knowing the thymolphthalein changes color at a pH of 9.0 (because we told you in the background info!), you might guess an answer of . . . oh . . . how about 9.0?

Test Tube Mania, page 96

1. The licorice root contains baking soda, which reacts with the acid to produce carbon dioxide. The carbon dioxide produces the foam.

2. You've added enough when the color doesn't change any further.

Sweet Tooth Favorite, page 101

2. Substances in a supersaturated solution will come out of solution and form crystals more readily than substances in a saturated solution.

BBQ Crystals, page 103

1. The crystals that are formed are ammonium salts.

2. A regular rock wouldn't be porous enough to soak up the solution.

Caffeine Crystals, page 105

3. The sublimed crystals are probably purer because any impurities would be left behind when the caffeine first evaporates.

Water Glass Garden, page 106

3. This is a chemical change. Entirely new crystals (the silicates) are formed.

Heating of Cobalt Chloride, page 107

3. The color change indicates this is a chemical change.

Conductivity, page 111
6. Yes. The human body is mostly water, but if that water doesn't contain ions, it won't conduct electricity.

Electrolysis of Water, page 114
1. The kids should notice bubbles forming at each electrode.
2, 3. Positive ions head for the negative electrode and negative ions head for the positive electrode. Opposite charges attract.

Sun Ribbon, page 116
2. The white coating is magnesium oxide, produced when the magnesium burns.